The Institute of Bi
Studies in Biolog

Endogenous Plant Growth Substances

by Thomas A. Hill M.A., Ph.D.

Lecturer in Botany, Wye College (University of London)

Edward Arnold

© Thomas A. Hill *1973*

First published 1973
by Edward Arnold (Publishers) Limited
25 Hill Street,
London, W1X 8LL

Reprinted 1975

Boards edition ISBN: 0 7131 2392 3
Paper edition ISBN: 0 7131 2393 1

All Rights Reserved. No part of this publication may be reproduced, stored in a retrieval system, or transmitted, in any form or by any means, electronic, mechanical, photocopying, recording or otherwise, without the prior permission of Edward Arnold (Publishers) Limited.

Printed in Great Britain by
The Camelot Press Ltd, Southampton

General Preface to the Series

It is no longer possible for one textbook to cover the whole field of Biology and to remain sufficiently up to date. At the same time students at school, and indeed those in their first year at universities, must be contemporary in their biological outlook and know where the most important developments are taking place.

The Biological Education Committee, set up jointly by the Royal Society and the Institute of Biology, is sponsoring, therefore, the production of a series of booklets dealing with limited biological topics in which recent progress has been most rapid and important.

A feature of the series is that the booklets indicate as clearly as possible the methods that have been employed in elucidating the problems with which they deal. Wherever appropriate there are suggestions for practical work for the student. To ensure that each booklet is kept up to date, comments and questions about the contents may be sent to the author or the Institute.

1972

INSTITUTE OF BIOLOGY
41 Queen's Gate
London, S.W.7

Preface

The study of the control of growth and differentiation in living organisms is one of the major preoccupations of current biological research, and this book looks at just one area of the subject. Many aspects of plant growth and development are controlled through the mediation of the endogenous plant growth substances or plant hormones, and research on these substances has a fairly long history. Developments from such studies have included the discovery of the many types of synthetic chemical with which we can control plant growth artificially in agriculture and horticulture, though these fall largely outside the scope of this particular book, in which I have tried to outline the present state of our knowledge of the plant hormones and to show some of the ways in which the many fascinating problems of their study have been approached.

I am grateful to the authors and publishers who have given their permission for the use of previously published illustrations, and to all those who have helped, knowingly or unknowingly, in the preparation of the book. The gestation period of these pages included part of a period of sabbatical leave which was spent working in the laboratory of the great French plant physiologist J. P. Nitsch, and the book is dedicated in respect and affection to his memory, and also to all his colleagues in gratitude for their stimulating and friendly hospitality.

Wye
1972

T.A.H.

Contents

Introduction 1

Plants normally grow and develop in an orderly, organized way. This is one of the many characteristics they share with all living organisms. Such orderly growth and development may be affected in many ways by the environment, which in some cases plays a very important part in controlling or triggering off various patterns of development. For example, some plants only produce flowers if the daily light period exceeds a certain critical length and some will only do so if they have been exposed to cold at an earlier stage in their development. In a general way, however, it is obvious that the essential features of growth and development are built into the genetic constitution of the plant and are thus controlled from within.

The mechanisms by which the internal control of growth is achieved in plants are many and complex and there is a tendency for each plant physiologist to look at such mechanisms from the point of view of his own area of specialization. This can be dangerous because the one thing which is certain is the very highly integrated nature of the controlling processes. Nonetheless, there is no harm in concentrating attention on one aspect of the subject at a time if it is always remembered that this is what one is doing. It is this philosophy which provides the justification for this book.

One of the important growth-controlling systems in plants is provided by the so-called 'plant growth substances' or 'plant hormones'. The term 'hormone' is, of course, used by animal physiologists too, and indeed the concept of such chemical messengers in the body was first developed by them in the early part of this century. Animal hormones differ from plant hormones in a number of ways, of which perhaps the most important is that in animals they are produced in specific organs or glands and often have very highly specific effects. Plant hormones, on the other hand, though they may be produced in fairly restricted regions of plants, are manufactured by unspecialized cells and frequently have many different effects upon the plant, depending upon other circumstances.

It will be useful to start with a definition of the terms 'plant growth substance' or 'plant hormone' (the two expressions will be used synonymously in this book):

> 'A plant growth substance (or plant hormone) is an organic substance which is produced within a plant and which will at low concentrations promote, inhibit or qualitatively modify growth, usually at a site other than its place of origin. Its effect does not depend upon its calorific value nor its content of essential elements.'

Not all plant hormones always fit this definition exactly, but it provides a useful working description of the substances we shall be discussing.

An immediate problem which arises is that there are many substances known which, when applied to plants, have effects which closely resemble those of the plant hormones. Some of these are actual chemical analogues of the endogenous plant hormones and some are not, but all can be synthesized in the laboratory. Such chemicals are very important in many types of research and some of them will be mentioned later. For example, some of the so-called hormone weed-killers fall into this category. Such substances are usually called 'plant growth regulators' or, more simply, 'plant regulators'. In many cases plant regulators of various kinds have been successfully used in the study of processes controlled internally by plant hormones.

One of the major difficulties of the study of endogenous plant growth substances is that the quantities of these chemicals present in plants are always very small indeed by the standards of normal techniques of chemical analysis. This problem is discussed in detail in Chapter 2. Another problem in considering the roles of endogenous plant hormones is that a very great deal of the relevant work is based on what is really circumstantial evidence derived from experiments in which the chemical, or a close relative of it, is applied to the appropriate plant from the outside. The reasoning in such cases is essentially something like this:

(a) We know that a substance *X*, or one very like it, occurs in a certain plant.
(b) We have a supply of substance *Y*, which is very similar to substance *X*.
(c) When applied to the relevant plant, substance *Y* causes a specific response (for example, stem elongation).
(d) Therefore it is likely or possible that substance *X* has a role in controlling stem elongation in this plant.

This is an over-simplification, of course, but there is no doubt that much of our understanding of plant growth substances depends on evidence which is at least partly of this kind. This is perfectly acceptable provided that we constantly bear in mind the possible sources of error in such arguments, and eventually try to confirm the conclusions by more direct methods.

Before going any further it will be useful to try to define the main groups of substances with which the rest of this book will be concerned. Not every point in the definitions will be clear at this stage, but it is important that they should be presented now and it may be useful to reread the definitions later on when some of the more detailed information about the substances has been covered (Chapters 3 and 4).

Five types of groups of substances will be discussed.

(a) AUXINS These are substances chemically related to indole acetic acid (IAA) (Fig. 1.1), which itself appears to be the principal auxin of many plants. They are characterized by their ability to promote growth in certain biological tests involving the use of excised parts of plants freed as far as possible from their own endogenous auxins. Many indole compounds have

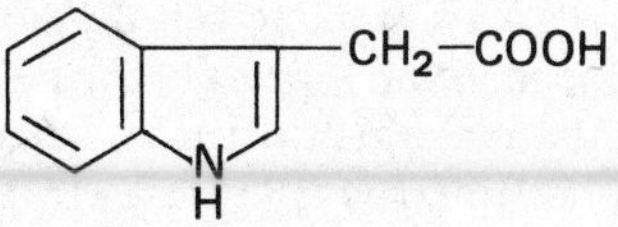

$$\text{(indole ring)}-CH_2-COOH$$

Fig. 1-1 Indole acetic acid (IAA).

IAA-like effects, as do many synthetic growth regulators which are not based on the indole framework. A well-known example of one such molecule is 2,4-dichlorophenoxy-acetic acid (2,4-D) which is one of the constituents of many weedkillers (Fig. 1.2).

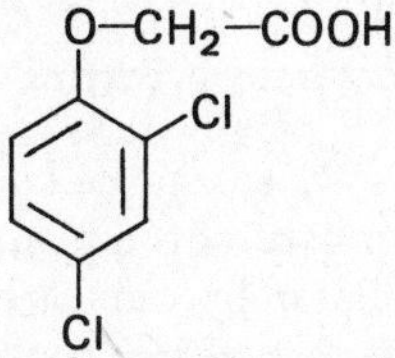

$$O-CH_2-COOH$$

Fig. 1-2 2,4-dichlorophenoxy-acetic acid (2,4-D).

(b) GIBBERELLINS These are substances chemically related to gibberellic acid (usually abbreviated to GA_3), which is a metabolic product of the fungus *Gibberella fujikuroi* and can be obtained from the liquid medium in which the fungus has been cultured. The gibberellin molecule is based on a gibbane skeleton (Fig. 1.3).

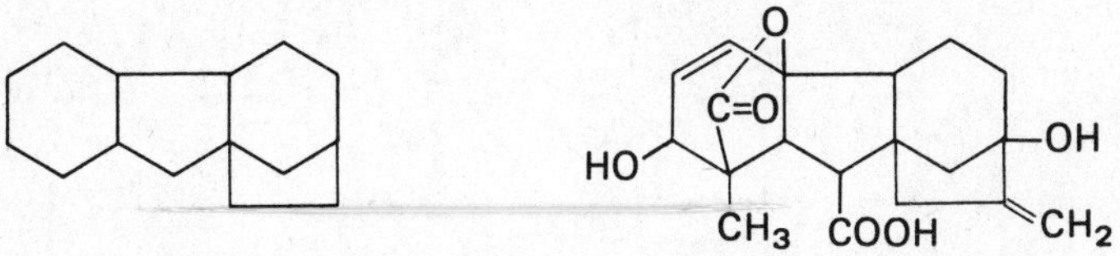

Fig. 1-3 The gibbane carbon skeleton (left) which is the basis of the known gibberellins, with the formula of gibberellic acid (gibberellin A_3, GA_3) (right).

Many different gibberellins have been found in plants and have been chemically characterized, and to a greater or lesser extent they all share the ability to cause stem elongation when applied to intact plants in the light; certain genetic dwarf strains of maize and peas are particularly sensitive test plants. Many cases exist where substances with properties resembling the known gibberellins have been extracted from plants but not in quantities permitting exact chemical identification. Such substances are spoken of simply as 'gibberellin-like substances'.

(c) CYTOKININS These are substances which are derivatives of the purine adenine (which is well known as one of the nitrogenous bases in the molecules of the nucleic acids DNA and RNA). They are characterized by their ability to interact with IAA to promote cell division in cultures of plant cells grown on artificial media, and especially by their property of affecting the patterns of differentiation which occur in such cultures. They have many other properties but this one is critical. An example of a naturally occurring cytokinin is zeatin, which has been obtained from maize grains (Fig. 1.4).

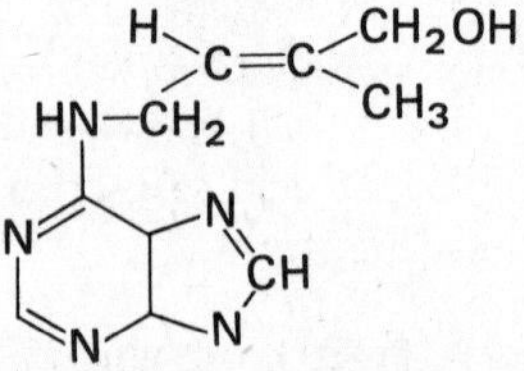

Fig. 1-4 The naturally occurring cytokinin zeatin.

The best-known cytokinin is a substance known as kinetin, but this has not so far been found to occur naturally in plants and should thus be classified as a plant growth regulator by our earlier definition. As with the gibberellins there are many cases of plant extracts containing cytokinin-like substances which are so called simply on the basis of their properties.

(d) INHIBITORS There are many substances present in plant cells which will, under some conditions, inhibit certain plant processes. Notable amongst these are phenolic compounds (see LEOPOLD, 1964, Chapter 9). However, the substances most able to be considered as inhibitors in the hormonal sense of our definition are those similar in structure and properties to abscisic acid (ABA) (Fig. 1.5). This substance is characterized by its

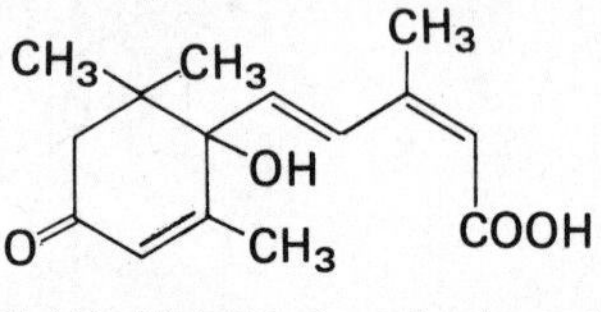

Fig. 1-5 Abscisic acid (ABA). This has also been known as Abscisin II and Dormin.

ability to inhibit many growth phenomena in plants, but perhaps especially by its association with bud dormancy in woody plants and with the abscission of leaves in the cotton plant.

(e) ETHYLENE (Fig. 1.6) This simple substance has long been known to affect the growth of plants, and since the development of sensitive techniques for detecting it and measuring its concentration it has become clear that it plays an important part in many plant growth responses. For ex-

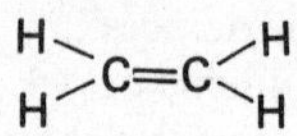

Fig. 1-6 Ethylene.

ample, it seems to be involved in many auxin-induced growth responses and it plays a part in leaf senescence, abscission and in the ripening of some fruits.

With the exception of ethylene it can be seen that we have had to define the groups of substances above largely in terms of their properties. Indeed in most cases they are defined in terms of what they will do to plants when applied exogenously. The reasons for this are largely those of convenience and historical accident. A large part of what we know about the plant hormones has been learned from experiments in which they were applied exogenously to plants (or parts of plants) and it is simpler to define them in terms of these effects than in any other way. This will become clearer as we look at the properties of the hormones in more detail and consider the relation between external application and endogenous role in the control of growth.

The fact that only five groups of plant growth substances are to be discussed does not necessarily imply that there are no other groups of plant hormones to be discovered, a point discussed further in Chapter 8.

A very short sketch of how the plant growth substances came to be discovered will now be useful. Fuller accounts of these studies can be found easily elsewhere, e.g. WAREING and PHILLIPS (1970); PHILLIPS (1971); WILKINS (1969); PRATT and GOESCHL (1969); ADDICOTT and LYON (1969).

1.1 Auxins

Figure 1.7 summarizes some of the crucial stages in the discovery of auxins.

Darwin's classical experiments with grass seedlings exposed to unilateral light showed that phototropic curvature was caused by a response in one part of the organ to a stimulus received elsewhere, and Boysen Jensen's work showed that the mediator of the response must be a chemical which moved in the plant. Paàl showed that uneven growth under unilateral light could be simulated by supplying the chemical stimulus from the apex unevenly. Went was able to isolate the chemical messenger by allowing it to diffuse into agar which was then placed asymmetrically on the decapitated seedling and which caused it to bend. The subsequent discovery of high levels of auxin activity in human urine and the culture filtrates of certain fungi greatly helped the study of the chemical nature of the substance, and indole acetic acid has now been isolated and chemically characterized from a wide variety of plants. Other indole compounds also occur in plants, but it is probable that their auxin-like activity is due to their conversion in the plant to IAA. IAA occurs not only as the free acid in plants, but is also often associated with other molecules, occurring bound to proteins or conjugated with amino acids or sugars (see THIMANN, 1969). A fascinating account of the very early work on auxins, in which the references in Fig. 1.7 may be found, is that of WENT and THIMANN (1937).

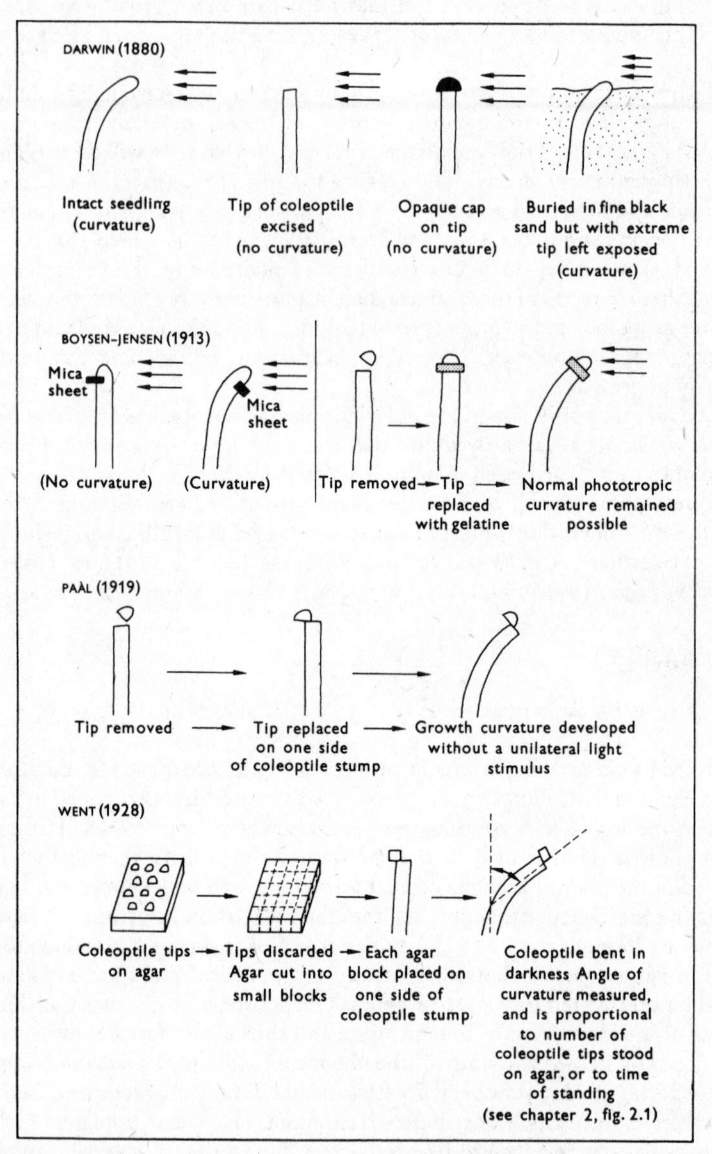

Fig. 1-7 A diagrammatic summary of some important steps in the discovery of auxins. References will be found in WENT and THIMANN (1937). Triple arrows indicate the direction of unilateral light, and all the experiments shown were done using grass seedling coleoptiles. (Modified from WAREING and PHILLIPS, 1970).

1.2 Gibberellins

The fungus *Gibberella fujikuroi* causes a disease of rice known in Japan as 'Bakanae' or 'Foolish Seedling'. The name derives from the appearance of infected plants, which are taller than healthy ones. In 1926 Kurosawa showed that this particular symptom could be duplicated by treating plants with the culture filtrate of the fungus. In the following thirty years much effort in Japan was devoted to trying to identify and characterize the causal chemical and to describe the physiological responses of plants to it. Much of this early work is in Japanese, but good accounts of it are available in English (e.g. STOWE *et al.*, 1961).

In the early 1950's workers in the U.S.A. and in England took up the study of gibberellins and as a result of the influx of new workers, the new chemical techniques by then available, and in some cases a certain amount of luck (see for example the paper by STOWE *et al.*, quoted above), the structure of GA_3 was worked out and a vast amount of work on its effects on plants began. The many different physiological effects of GA_3 on plants had led to predictions that a substance similar to it would be found in higher plants, and in fact the first evidence suggesting this was published in 1956 (WEST and PHINNEY, 1956; RADLEY, 1956). Since that time a large number of gibberellins has been isolated, purified and chemically characterized. Some of these are of fungal origin, some from higher plants and some (e.g. gibberellic acid itself) from both sources. There are also many reports of substances in plant extracts which are clearly gibberellins but which have not so far been chemically characterized as such. The known gibberellins are referred to by numbers which are approximately in chronological order of the discovery of the substances.

1.3 Cytokinins

Studies of plant cells grown under sterile conditions on synthetic nutrient media led to the discovery of this class of plant hormones. Skoog and his colleagues at Wisconsin worked with a strain of tobacco in which stem cells would grow in pure culture only if IAA were present. They showed that while cell division would take place in such cultures if vascular tissues were present, isolated pith cells responded only by cell enlargement. Cell division in such pith cultures could be induced by supplying malt extract or coconut milk. It was later shown that autoclaved DNA was highly active and from this work stemmed the isolation and characterization of the powerful cell-division-inducing substance kinetin. A good account of this early work is given by FOX (1969).

Kinetin was found to have many effects on plants other than the particular one described and the search for endogenous substances with similar properties in plants became very active. Although many plant extracts were soon found to contain substances with kinetin-like activity it was 1964 before the first naturally occurring cytokinin was finally characterized.

All the cytokinins so far discovered are derivatives of adenine, and since such derivatives are relatively easily synthesized much work has been possible on the relationship between structure and activity in cytokinins which has not so far been possible in the case of gibberellins.

1.4 Inhibitors

An account of many growth inhibitory substances in plants is given by LEOPOLD (1964) but from the point of view of hormone physiology the history of the most interesting inhibitor (known since 1967 as abscisic acid or ABA) only dates from about that time. The story of ABA is interesting because it is a striking example of several groups of workers reaching the same conclusion by different routes. One group in California, led by Carns and Addicott, was working on the problem of abscission in the cotton plant and isolated from cotton fruits an inhibitor whose level was associated both with premature abscission in young fruits and with the final abscission of mature ones. They published an account of the structure of this substance in 1965 and called it abscisin II.

A group of British workers led by Wareing at Aberystwyth, working at the same time on the control of dormancy in deciduous trees, isolated an inhibitor which seemed to be associated with this phenomenon. Chemical analysis showed that this inhibitor (which they called dormin) was in fact identical with abscisin II.

A third study, on an inhibitor associated with fruit abscission in lupin pods, culminated in the report by CORNFORTH *et al.* (1966) showing that this too was identical with abscisin II.

Subsequently an enormous amount of attention has been devoted to ABA, and to the study of related inhibitors in plants.

1.5 Ethylene

Ethylene has been known for many years to have an effect on plants. Early in this century it was shown that traces of it in illuminating gases in the laboratory affected geotropic behaviour of stems and roots and caused stunting and increased radial growth of pea stems. The production of ethylene by some ripening fruits and its effect on triggering off ripening in other fruits were also well established in the first part of the century. As early as 1935 there were suggestions that ethylene might be regarded as a hormone and the production of ethylene by various plants was shown at about the same time. However, right up to the recent past the difficulty of detecting and measuring the very minute quantities of ethylene involved in many plant responses has been so great that reliable interpretation of its hormonal role has been difficult. The advent of gas-solid chromatography (see Chapter 2) has revolutionized the study of the role of ethylene in plants and as a result we are at present in the middle of an explosion of new interest in ethylene as a plant hormone. Much of the recent work is reviewed by PRATT and GOESCHL (1969).

Methods 2

All the books which deal with plant growth substances contain a certain amount of detailed information about the methods which are used in their study (e.g. AUDUS, 1959, LEOPOLD, 1964; WAREING and PHILLIPS, 1970; PHILLIPS, 1971). Many of these methods are self-explanatory in their contexts, but it is essential here to look briefly at some of the experimental techniques which are in common use. Much of the work which has been done involves either studies on the responses of plants to applied plant growth substances or the separation, purification and assay of these substances from tissues. It is these last three aspects which are dealt with in this chapter.

One of the basic problems of the study of endogenous plant growth substances is the very small quantities which are present. The units in which quantities of plant hormones are expressed are very often microgrammes (μg), that is millionths of a gramme, and extracts often contain only small fractions of a microgramme. Concentrations of hormone solutions are normally expressed either in terms of molarity or in parts per million (ppm = $mg/dm^3 = \mu g/cm^3$). In terms of normal chemistry these quantities are extremely minute and are very difficult to envisage. Most normal chemical techniques are far too insensitive to detect and measure quantities of substances of this order and biological tests have to be employed instead once the growth substances have been extracted from the tissues containing them.

The methods used to extract and purify growth substances from plant tissues are very varied, and have tended to be developed on a rather *ad hoc* basis. For example, many of the early attempts to extract gibberellins from plants used ethyl ether as a principal solvent, solely because it had been used with success for the extraction of auxins in earlier years. A difficulty in selecting appropriate extraction procedures is that the substance being extracted may be of unknown chemical composition. It is thus impossible to be certain that the extraction procedures in use are not altering the structure and properties of the material. Some indole compounds are rather labile, especially under acid conditions, and serious losses can occur in the course of extraction unless adequate precautions are taken. Some gibberellin-like substances have been shown to change their properties after extraction if left for some time. In a good many cases one has no alternative but to put up with these inconvenient possibilities until techniques become available to overcome them, but it is important to bear in mind that such problems do exist, otherwise results may be misinterpreted.

2.1 Extraction procedures

The aim of all extraction procedures is to separate the plant growth substance from everything else with as little loss as possible. Auxins and gibberellins are usually extracted by one of two main types of technique, diffusion or solvent extraction.

2.1.1 *Diffusion techniques*

In this method the organ from which the substance is to be extracted is excised and placed with the cut surface on agar jelly, which may be replaced at intervals. Any substance moving from the tissue is subsequently extracted from the agar by other means.

The advantages of this technique are that while many substances present in the tissues do not readily diffuse into agar in this way, under appropriate conditions auxins and gibberellins do, so that the method combines extraction with partial purification. Went, in his early work, used the technique to demonstrate the diffusibility of auxin by cutting up the agar after diffusion and placing pieces asymmetrically on decapitated oat coleoptiles in the dark. The subsequent curvature of these coleoptiles gave a measure of the amount of auxin which had diffused into the agar (Fig. 2.1). A disadvantage of the diffusion method is that it is really only practicable for small quantities of tissue. A further problem is that there is no way of estimating the

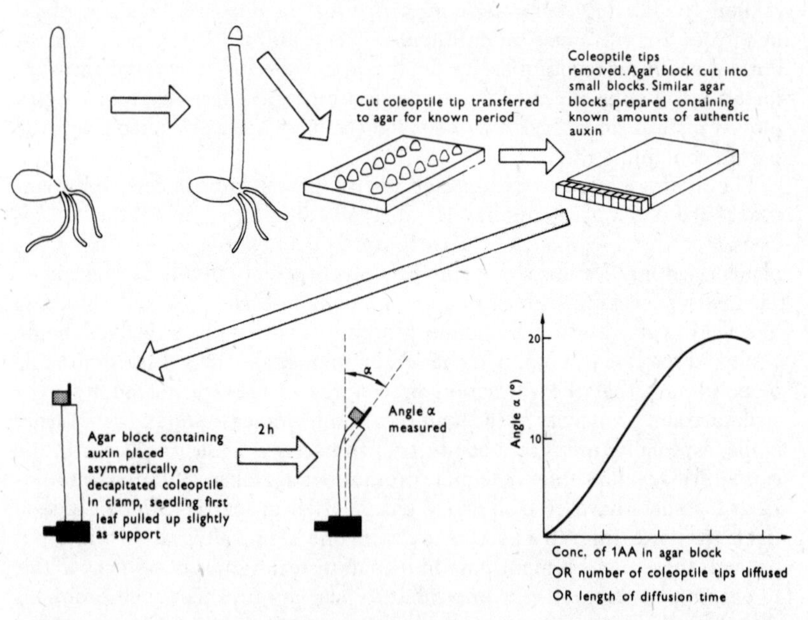

Fig. 2-1 An outline of the *Avena* Curvature Test for auxins, showing the relationship to Went's original quantitative studies referred to in Chapter 1. In the case illustrated the amount of auxin diffusing from the coleoptile tips could be estimated by reference to the standard curve prepared from the responses of coleoptiles to known concentrations of IAA.

losses of activity which may take place due to chemical action at the cut surface. On the other hand the technique has the advantage that it may be used to estimate the actual rate of production of substances by small pieces of tissue.

2.1.2 *Solvent extraction*

The principle employed here is that when a substance is shaken up with two immiscible solvents the proportion dissolved in each solvent will depend on the partition coefficient of the solute in the particular system. Since different substances have different partition coefficients, a series of transfers from one solvent to another can be used to 'leave behind' undesirable materials while purifying the required substance. In the case of IAA and many gibberellins and inhibitors which are acidic it is possible to

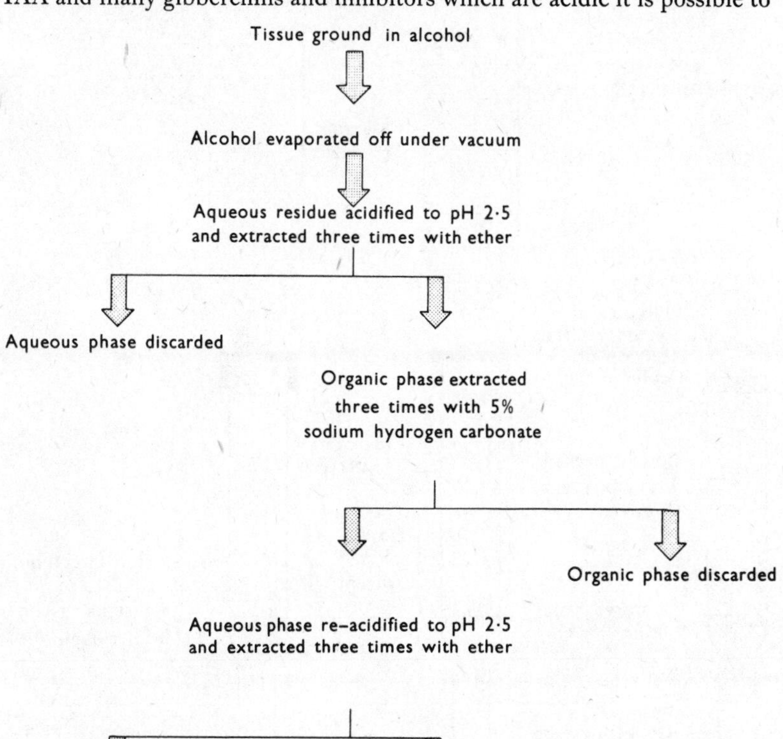

Fig. 2-2 Flow diagram illustrating an extraction procedure for acidic auxins in plant material.

alter the partition coefficient between an aqueous (polar) solvent and an organic (non-polar) solvent by adjusting the pH of the former. At low pH values acidic substances are less ionized and are soluble in non-polar solvents, while at high pH, in their ionized form, they dissolve much more readily in polar solvents. The flow diagram in Fig. 2.2 illustrates a typical procedure for the separation of an acidic substance from plant tissue.

The choice of solvents, pH values and other details depends on the substances being studied. In the case of gibberellins for example, ethyl acetate is a much more satisfactory solvent than ethyl ether. By appropriate modification solvent extraction techniques can be used also for substances which are neutral or basic rather than acidic.

Ethylene is unique amongst the plant growth substances in that it is

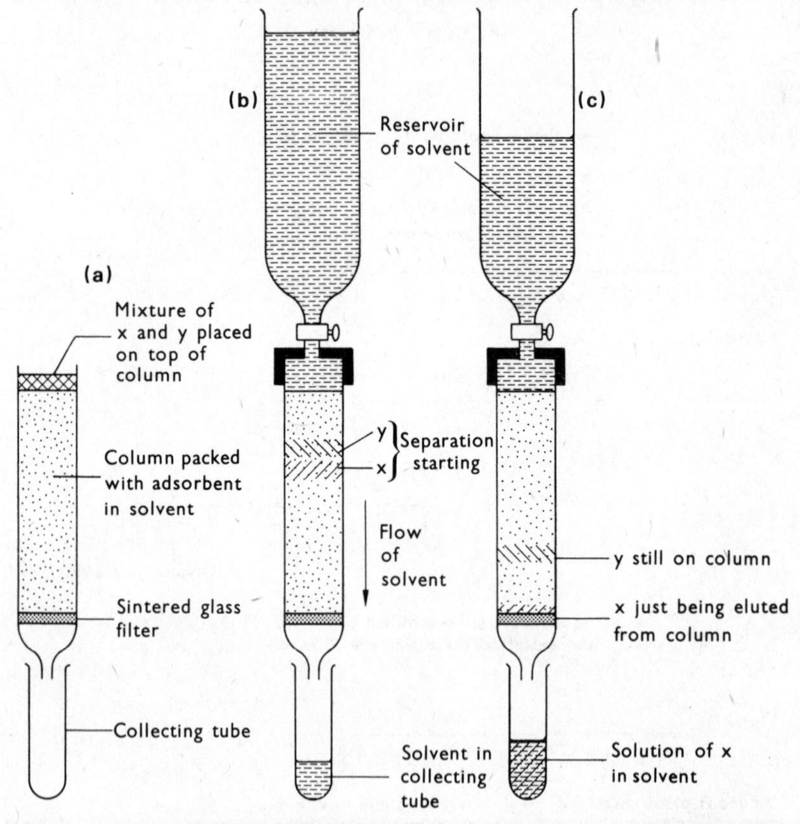

Fig. 2-3 Diagram illustrating the principle of column chromatography. (a) A mixture of two substances (x and y) is placed on top of the column. (b) The flow of solvent from the reservoir moves the substances down the column, x faster than y. (c) x is eluted from the column and collected while y is still on the way down.

gaseous and diffuses into the surrounding air. No extraction procedure for it is therefore necessary; all that is needed is an arrangement to prevent it from escaping from the experimenter altogether!

Once the growth substances have been extracted from the plant further purification is almost always necessary, and this is usually done by some form of chromatography. A vast range of chromatographic techniques is available suitable for substances of many kinds. These include chromatography on paper, on columns of suitable adsorbents and on glass plates spread with very thin films of adsorbents. In each case the materials being purified are moved along the supporting medium by some appropriate solvent and different substances are separated according to their solubility in the solvent or their affinity for the adsorbent, or both. (See Fig. 2.3.)

All these methods of extraction and purification depend absolutely on our ability to recognize and assay the substance separated, but even when adequate assay procedures are available unexpected problems can arise. MANN and JAWORSKI (1970) showed that there could be substantial losses of IAA by oxidation during paper chromatography even though stringent precautions were taken to avoid losses earlier in the extraction. Substances with gibberellin-like activity in some tests can arise directly from the solvents used in extraction procedures, a problem which can be overcome only by changing the procedures entirely or by the use of rigorous control experiments to keep a check on the extent to which this may be happening. This is the sort of problem which arises as a result of the minute amounts of hormonal substances needed to give responses in biological tests.

2.2 Biological assays

So much of the interpretation of the results of plant growth substance research rests on biological assays, as has already been implied, that it is essential to say something specifically about them.

A biological assay, also called 'bioassay', or 'biological test' is essentially any system in which the quantity of a substance applied can be estimated by measuring the size of an organism's response to the substance. A 'standard curve' is produced which relates the response of the organism to known doses of the substance and from this the dose corresponding to any particular size of response can be read off, enabling us to estimate unknown concentrations with reasonable ease. As far as plant growth substance bioassays are concerned there are a number of features which a good assay should possess. These include:

(i) *Sensitivity* A detectable response should be obtained from a very minute amount of the substance.

(ii) *Specificity* An assay which responds to more than one type of substance is much less useful than one which responds to only one type.

(iii) *Insensitivity to environment* Assays which give similar results in a fairly wide range of conditions are more useful than those which require very closely controlled conditions.

(iv) *Ease of execution* Some biological assays require great technical skill to give reliable results. Those which are easy to do are obviously preferable, other things being equal.
(v) *Rapidity* Some bioassays give a result in a few hours or a few days. Others take weeks or even months and this is obviously a grave disadvantage in many sorts of work.
(vi) *Availability of material* Assays using commonly available plants or cultivars are much more convenient than those needing special supplies of seed or plant material from other parts of the world.
(vii) *Linearity of response* The interpretation of the results of bioassays is made much easier if the relationship between the dose of the substance and the response of the test system is linear, and also if the linear portion of the response curve covers a wide range of doses.

In the case illustrated in Fig. 2.4, bioassay B would be preferable to bioassay A on both these grounds. In the case of most plant growth sub-

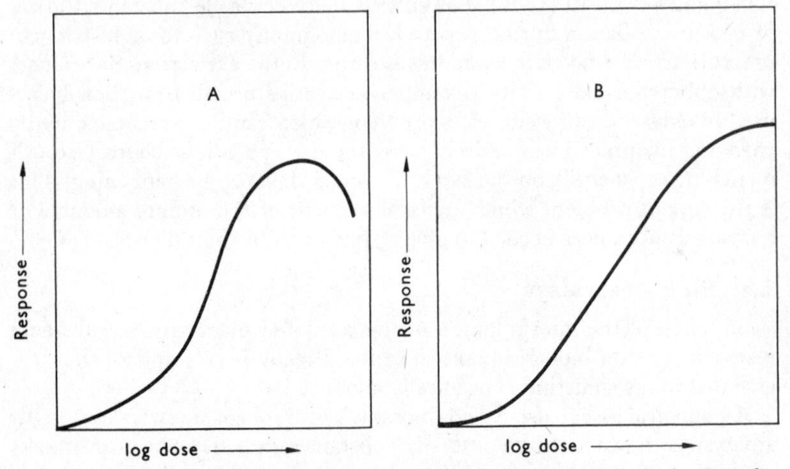

Fig. 2-4 Hypothetical log dose/response curves for two biological assays for a plant growth substance.

stance bioassays the response is more usefully plotted against the logarithm of the concentration, as is shown in Fig. 2.4. A result of this is that, although the range of concentrations over which the response is linear may be very large, it is not possible to read off the dose corresponding to a given response directly unless the data is plotted on logarithmic paper. In other cases a small calculation is required.

There is probably no bioassay which fulfils all these criteria, but they form a useful basis on which to judge an unfamiliar test. The results of many different types of biological assay are quoted in later chapters, but it may be useful here to summarize one or two which are in common use for the various groups of growth substances with which we are concerned.

2.2.1 *Indole auxins*

(a) THE OAT OR WHEAT COLEOPTILE STRAIGHT GROWTH TEST In this test oat or wheat seedlings are grown in the dark (with a short exposure to red light in the case of oats, to suppress growth of the mesocotyl) until the coleoptiles are 2–3 cm long. The apical 3–4 mm of the coleoptiles are removed and a segment of standard length cut from the remaining part and floated on the test solution. The increase in length of this after a period (usually 24 h) in the dark is taken as a measure of auxin activity.

(b) THE WENT PEA CURVATURE TEST This test, which is useful for rapid screening of compounds for auxin-like activity, is carried out by floating sections of the stems of dark-grown pea seedlings for 24 h on solutions of the test substances. The sections are slit lengthways for about $\frac{2}{3}$ of their length before use, and the degree of bending of the two arms so formed gives a rapid visual estimate of the activity of substances compared with standard solutions.

2.2.2 *Gibberellins*

There are a great many biological assays for gibberellins, and these have been reviewed by BAILISS and HILL (1971).

(a) DWARF PLANT BIOASSAYS The dwarf pea bioassay in which stem length is measured, and the dwarf maize assay in which the length of the leaf sheaths of gibberellin-treated dwarf maize plants gives a measure of the amount of gibberellin used, are two of the most widely used assays for these substances. Not all dwarf cultivars, however, show this type of response to gibberellins.

(b) BARLEY ENDOSPERM BIOASSAY This test depends on the gibberellin-induced increase of the enzyme α-amylase in barley grains cut in half to remove the embryo. In one version of the test embryoless half grains are placed in the test solutions and the reducing sugars released from the starchy endosperm are assayed chemically after one or two days. An alternative method is to grind up the half-grains after the appropriate incubation period, extract the enzyme and assay this by measuring its effect in hydrolysing a starch solution of known concentration. Both methods are being increasingly used since this test is sensitive, rapid, and is well suited to processing large numbers of samples in a short time. It is also often favoured because the effect of gibberellic acid on enzyme synthesis is thought to be a very direct one, whereas tests involving the growth of whole plants are probably measuring the gibberellin effect in a rather indirect way, a matter discussed further in Chapter 7.

2.2.3 *Cytokinins*

(a) TOBACCO CALLUS TEST Cytokinins have the specific property of inducing the growth of cultures of callus tissue derived from tobacco pith cells. Small pieces of tissue are placed on appropriate agar media containing the test substance and incubated for a period of some weeks. The final weight of the culture is a function of cytokinin concentration. A disadvantage of this test is the length of time needed to obtain results.

(b) LEAF SENESCENCE TESTS In some plants cytokinins delay the loss of chlorophyll from leaf sections or discs floated on solutions in the dark. The amount of chlorophyll remaining in radish leaf discs or wheat leaf sections after 3–4 days can be assayed spectrophotometrically, and is a function of cytokinin concentration. The test is not as specific as the callus test, but is much easier to carry out.

2.2.4 *Inhibitors*

It is not easy to find good biological tests for endogenous growth inhibitors in plants, and there are none that are specific to known groups of substances. In order to assay inhibitors successfully it is necessary to use a test system which is growing rapidly, and many workers have used modifications of the wheat or oat coleoptile test for auxins in which the coleoptile is caused to grow rapidly by the addition of auxins or of sugar and other substances to the incubation medium. The degree of inhibition of this induced growth can then be measured and related to inhibitor concentration. Other workers have used the inhibition of seed germination or of hypocotyl growth in young seedlings, but it remains true that mainly because of their lack of specificity, inhibitor assays are less satisfactory than those for growth promoting substances.

2.2.5 *Ethylene*

There is no satisfactory bioassay for ethylene, though the growth response of pea seedlings was used for many years (see Chapter 4). Fortunately ethylene can now be readily detected and assayed by gas-solid chromatography (see below).

2.3 Physico-chemical methods

The physico-chemical methods used in the study of plant growth substances are all of relatively recent origin, but are likely to become increasingly important as they become more and more sensitive and flexible. It is useful here to mention two main types of technique, one of very varied application, the other essentially analytical.

2.3.1 *The use of radioactive tracers*

Where plant growth substances can be synthesized it is possible to incorporate into the molecule radioactive atoms at almost any desired position. For example, extensive use has been made of IAA labelled with the carbon isotope ^{14}C in studies on the movement of IAA in plants. The advantage of work with radioactive substances is that extremely minute amounts can be detected. A problem which does arise, however, is that it is always necessary to be sure that the material whose radioactivity is being measured is the same as the material originally introduced into the plant. This has to be done by combining examination of the pattern of radioactivity in the plant following the application of the isotopically labelled substance with orthodox extraction of the radioactive regions. Where, for example, extraction and chromatography show that all the radioactivity in a particular region is associated with a single substance with chromatographic behaviour similar to the material originally applied, it can be reasonably assumed that the substance is not being altered in the course of the experiment. On the other hand, a great advantage of the use of radioactive substances is that if they are broken down in the plant it is often possible to trace the pathways of breakdown by studies on the radioactive intermediates which are to be found in extracts.

Up to the present it has been possible to synthesize radioactive samples of IAA, of all synthetic auxins, of synthetic cytokinins, of some inhibitors and of ethylene. Some radioactive gibberellins have been produced by allowing the fungus *Gibberella fujikuroi* to synthesize them in the presence of radioactive substances which are known precursors of gibberellins.

2.3.2 *Gas–liquid chromatography and mass spectrometry*

A recently developed technique, gas–liquid chromatography (GLC) is applicable to any substance of which a volatile derivative can be made. In the case of gibberellins, for example, the methyl esters can be prepared; these are separated on a column containing an inert support such as a diatomaceous earth which is coated with a substance (called the stationary phase) in which the ester is soluble. A stream of nitrogen is passed through the column and carries the volatile components of the mixture through the column, from which they emerge one at a time according to their various partition coefficients between the stationary phase and the carrier gas. The substances can be detected at the column outlet by a variety of extremely sensitive techniques from which electrical currents can be derived. After amplification these can be recorded graphically on a moving chart, and since for a given substance in a given experimental system the time from the injection of the sample at one end of the column to its emergence at the other (the retention time) is characteristic, this can be used for identification purposes. The size of response recorded is also a measure of the quantity of the substance present. This technique is becoming very useful in studies on gibberellins, and in particular it has enormous analytical power when combined with the technique of mass spectrometry (see for example MACMILLAN and PRYCE, 1968). In such systems the gas chromatograph is connected to a mass spectrometer, and this enables both the molecular weight and a great deal of detailed information to be obtained automatically about the chemical structure of any substance detected by the gas chromatograph. In comparison with the immensely laborious task of separating enough of an individual substance for chemical analysis this technique has very obvious attractions for those laboratories with the necessary very expensive equipment.

In the special case of ethylene referred to earlier this technique is properly referred to as gas-solid chromatography, since no stationary phase is needed. In practice the general term 'gas chromatography' tends to be used for both GLC and GSC techniques.

In all plant growth substance work the problem of technique looms large, because the interpretation of results in terms of explanations of plant growth may depend on a fairly exact appreciation of the sometimes serious limitations of the techniques used. A great deal of work in fact goes into the devising of new techniques and the refinement of old ones, but it is possible to become so transfixed by problems of technique that no conclusions are ever drawn at all. This error is probably much more rare than its converse, the drawing of conclusions on a thoroughly unsatisfactory experimental basis.

The hormonal control of growth I. Auxins and gibberellins 3

3.1 Introduction

In this chapter and the next we shall consider in rather more detail some of the evidence which associates plant growth substances with the control of growth.

It would take up an enormous amount of space to describe fully the many effects the plant hormones have on the growth and development of plants. Some of the more important of these effects have therefore been summarized in Table 1. In this Table a fairly large number of the effects caused by exogenously applied plant growth substances are listed in the first column, and the remaining five columns contain information about the involvement of each group of substances in these effects. One advantage of giving the information in this form is that it may be seen at a glance that there are many processes which are affected by more than one type of plant growth substance.

A symbol (*) has been used in the Table in a number of places. This indicates that there is either clear or strong presumptive evidence that the process referred to is controlled endogenously by the relevant group of substances. Both the symbol and the information in the body of the Table should be interpreted with caution, since in some cases the responses referred to may only occur in a few plant species (or, of course, only a few species may have been investigated). The evidence for the involvement of endogenous hormones may be fragmentary, and in some cases the symbol has been omitted for this reason, even though endogenous hormones may very well be involved. Errors and omissions can only be blamed on the author, and the reader may like to amuse himself by trying to find some of these.

Entries in capitals in Table 1 are an indication that effects are particularly characteristic of the relevant substance and have been fairly extensively studied as such.

In this chapter and the next, the five groups of hormones under discussion are all dealt with under the same headings. These are:

(i) *Effects on plants* This includes mainly those effects not already listed in Table 1.

(ii) *Evidence for the occurrence of the hormone in plants*

(iii) *Evidence for the hormonal role of the substance* Here an attempt is made to show what sort of evidence enables us to regard the substance as a hormone in the sense of the definition in Chapter 1.

Table 1 Some effects of exogenous plant growth substances on plants. Effects in capitals are of significance as characteristic of particular substances. The symbol * indicates that there is evidence for the role of the relevant endogenous substance in the control of the process referred to.

	Effect produced by exogenous application of growth substance	*Auxins*	*Gibberellins*	*Cytokinins*	*Abscisic acid*	*Ethylene*
1	Affects tropic responses	Yes *	Yes*	No	Yes: endogenous inhibitors may be involved in some aspects *	Yes. Associated with auxin relations of the tissue *
2	Active in *Avena* curvature test	Yes *	No	No	Not tested	Not applicable
3	Affects growth in excised segments of wheat or oat coleoptiles	YES, PROMOTES UNDER APPROPRIATE CONDITIONS *	Promotes in some cases but less typically than auxins	Promotes in one or two specific cases	Inhibits	Pre-treatment may enhance the auxin response, but normally ethylene inhibits longitudinal growth but causes radial expansion
4	Causes cell enlargement in tissue cultures (i.e. in largely non-polar growth)	Yes (some cases)	Yes in some cases, but not usually essential	Yes, but also causes cell division in interaction with auxin	No	No
5	Controls differentiation in tissue cultures	YES, WITH CYTOKININS	No	YES, WITH AUXINS	No	No
6	Stimulates root initiation in cuttings	Yes *	No. Often inhibits	Variable response	Has been reported in some cases	Yes
7	Inhibits root growth	YES *	No. Occasionally promotes	Little studied	Yes	Yes
8	Stimulates cambial division	Yes *	Yes	Yes, in some cases, but little studied	May inhibit *	No, but causes radial expansion of stem cells in some legume seedlings
9	Affects xylem differentiation	Yes *	Yes	Yes, in some cases but little studied	Yes, in some cases, but little studied	Not studied
10	Affects abscission of leaves or fruits	Yes, largely through effect on ethylene production*	Not directly	Probably, through effect on delay of senescence	Yes. Accelerates* this in some plants (e.g. cotton, lupin)	YES *
11	Promotes fruit growth	Yes *	Yes *	Yes, in some cases	No	No

12	Affects stem growth in intact plants in the light	No	YES, PROMOTES *	No	May inhibit with repeated application	Inhibits
13	Promotes stem elongation and flowering in rosette biennials	No	YES *	Yes, but only known in one plant	No	No
14	Promotes flowering in other types of plant	Yes, in pineapple due to ethylene production	To some extent; inhibits flower initiation in some woody plants, e.g. apple	No	Inhibits flowering in some Long Day plants under inductive daylengths. Effects on Short Day plants variable *	YES, IN PINEAPPLE
15	Reverses some types of genetic dwarfism	No	YES *	No	No	No
16	Releases vegetative buds from dormancy	No	Yes *	Some cases* reported; may be associated with bud growth in some other cases, e.g. effect on lateral buds	NO. INDUCES DORMANCY IN SOME WOODY SHOOTS*	Yes, in some cases
17	Causes germination in light-requiring seeds	No	Yes	Yes, in some cases	No. Inhibits	Yes
18	Accelerates seed germination in general	No	Yes *	No	NO. TENDS TO INHIBIT *	Yes, in some cases *
19	Causes synthesis of α-amylase in cereal grains	No	Yes *	Has been reported	No. Inhibits this. (Effect reversed by kinetin)	No, but may increase the release of enzyme already present
20	Causes expansion of discs cut from leaves	No	Yes, in some cases	Yes, in some cases	No	No
21	Has an effect in maintaining or breaking apical dominance	YES. APICAL APPLICATION MAINTAINS APICAL DOMINANCE*	Yes, normally increases apical dominance*	Yes: often breaks dominance if applied to lateral buds*	Not studied	Yes
22	Inhibits protein and chlorophyll breakdown in senescence	Some cases reported	Yes, in some cases	YES, IN SOME CASES	No. Accelerates this, notably in leaf discs	No, tends to accelerate senescence*
23	Promotes climacteric respiratory rise in ripening fruit	Not studied, but probably due to ethylene production	No	No	No	YES *

(iv) *Modes of action* The whole of Chapter 7 is devoted to a discussion of possible modes of action of gibberrellins and auxins, and this section is consequently not relevant in Chapter 3. Brief reference is made to the modes of action of cytokinins, ABA and ethylene in Chapter 4, but these short sections might be better understood if read after, rather than before, Chapter 7.

3.2 Auxins

3.2.1 *Effects of auxins on plants*

The more important effects of auxins on plants are listed in Table 1. One point which is of interest in relation to these effects however, is that when IAA is applied exogenously to plants it is rarely toxic. Although at high concentrations it may inhibit many processes and cause toxic symptoms for a time these usually disappear as the excess auxin is destroyed by an enzyme complex usually referred to as 'IAA oxidase'. Some of the synthetic substances which have auxin-like properties at low concentrations are herbicidal at high concentrations because such destruction does not take place.

3.2.2 *Evidence for the occurrence of auxins in plants*

The evidence for the natural, widespread occurrence of IAA as the principal auxin in plants is now overwhelming. It has been chemically identified from a wide variety of sources and there is chromatographic evidence, not yet backed up by chemical identification, for its occurrence in many others.

It was thought for many years that a principal precursor of IAA in plants was the amino acid tryptophan (Fig. 3.1) and conversion of this to

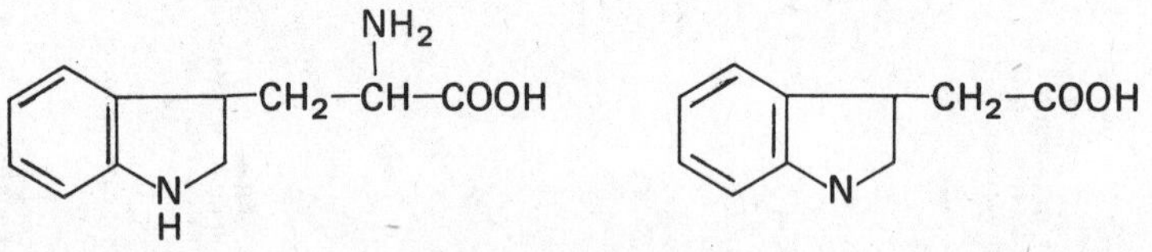

NH_2 | $-CH_2-CH-COOH$ N H $-CH_2-COOH$ N

Fig. 3-1 Formulae of tryptophan (left) and IAA, to show their close similarity.

IAA has often been reported. Recently, however, it has been shown that in some plants such conversions appear to be carried out only by bacteria which are present in the tissues. Plants grown under completely sterile conditions failed to produce IAA from tryptophan and also contained lower concentrations of IAA than non-sterile controls (LIBBERT *et al.*, 1969), and uncertainty remains as to the precise route and source of IAA synthesis in many tissues.

IAA occurs in plants both as the free acid and conjugated with such substances as glucose, glutamic and aspartic acids. It also occurs closely associated with proteins from which it may be released by protein hydrolysing enzymes. The physiological role of some of these so-called 'bound' forms of IAA is not yet known, though some may act as a reservoir of the substance, being formed when large amounts are present in the cell and broken down when IAA levels are low.

Some other indole compounds which occur in plants and whose auxin-like effects appear to be due to enzymic conversion to IAA are shown in Fig. 3.2.

R = (indol-3-yl ring, N–H)

Formula	*Name*	*Source*
$R \cdot CH_2{-}CHO$	Indole acetaldehyde	Etiolated seedlings and other sources
$R \cdot CH_2{-}C({=}O){-}COOH$	Indole pyruvic acid	Maize grains and other sources
$R \cdot CH_2{-}CN$	Indole acetonitrile	Many members of the Cruciferae. Probably does not occur free, but is formed from glucobrassicin (below) by enzymic hydrolysis
$R \cdot CH_2{-}C({=}N{-}O{-}HSO_3)(S{-}C_6H_{11}O_5)$	Glucobrassicin	Many members of the Cruciferae
$R \cdot CH_2{-}CH_2OH$	Indole ethanol	Cucumber seedlings

Fig. 3-2 Some derivatives of IAA found in plants. All have been shown to be active in some bioassays for IAA, and activity is probably due to their conversion to this substance by the plant.

3.2.3 *Evidence for the hormonal role of auxins*

The best known evidence for the hormonal role of IAA is the part it plays in phototropic and geotropic curvature by moving in the tissues in response to the stimulus. It seems clear that IAA plays a fundamental role in stem elongation phenomena in general. In a number of cases excised segments of stems, washed free of much of their endogenous auxin, will increase in length in response to exogenous IAA, but exogenous IAA rarely causes an increase in length of intact stems. This can best be explained on the hypothesis that in such stems endogenous auxin is usually at or near an optimal level. Treatment with GA_3, which does cause growth in intact stems, also frequently causes an increase in endogenous IAA level in the tissues. In general, changes in stem growth rate are often closely accompanied by changes in concentration of endogenous auxin, and a similar phenomenon in expanding fronds of the fern *Osmunda* is illustrated in Fig. 3.3. Such evidence on its own is, of course, largely circumstantial, since frequently one cannot say whether the auxin changes observed are a cause or a consequence of the related growth changes.

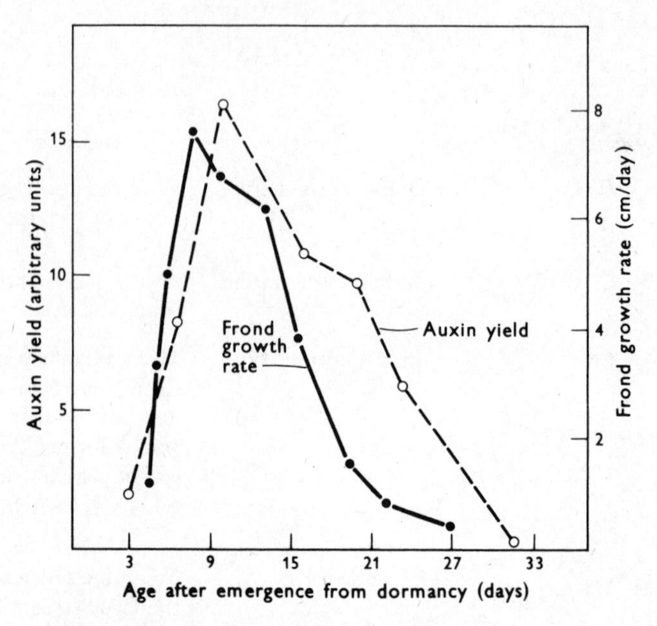

Fig. 3-3 Relationship between diffusible auxin and growth in expanding fronds of the fern *Osmunda cinnamomea*. (After STEEVES and BRIGGS, 1960.)

Exogenously applied IAA initiates cell division in the cambium and there is good evidence that IAA is involved naturally in this phenomenon. In the spring, when deciduous woody shoots start to grow, there is a sharp increase in auxin production by the apical buds, and this auxin moves down the shoot towards the base of the plant. As it moves cambial division starts near the shoot apex and also spreads downwards. If the apical bud is removed no such cambial division takes place; replacement of the excised bud by lanolin paste containing IAA re-introduces downward movement of the hormone and cambial division begins as usual. This evidence for the hormonal role of IAA is not weakened by the fact that other hormones can also be shown to have effects on division of cambial cells. Not only cell division but also the pattern of differentiation of the cells formed from the cambium appears to be controlled hormonally.

The classical work of the French physiologist J. P. Nitsch on the growth of the strawberry provides a final example of the hormonal role of IAA. Nitsch showed that the developing achenes on the strawberry were a rich source of IAA and that if they were removed the 'fruit' did not swell. Replacement of the developing achenes by synthetic auxins in lanolin caused virtually normal receptacle swelling, giving a smooth, seedless 'fruit' (NITSCH, 1950; PHILLIPS, 1971). Fruit shape was affected by the

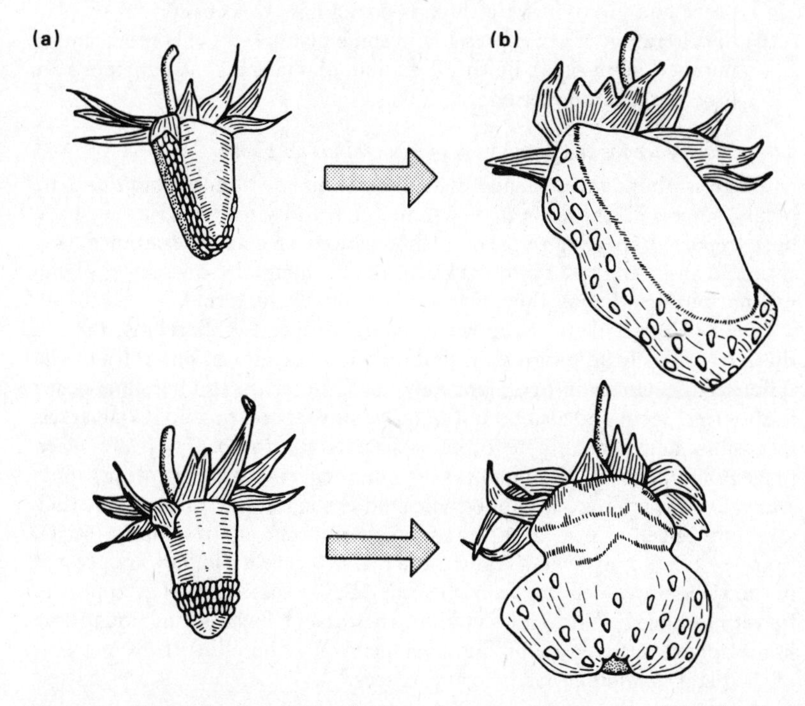

Fig. 3-4 The influence of developing achenes on the growth of the 'fruit' of strawberry. (a) Two young fruits with all but three rows of young achenes removed. (b) The same two fruits showing the consequent abnormal growth of the receptacle. (After NITSCH, 1950.)

removal of different achenes (Fig. 3.4) and it is a generally observed phenomenon that failure of seed development in many fruits leads to uneven growth and mis-shapen fruit. If a lop-sided apple is cut in half transversely it will frequently be found that the seeds in the part of the ovary nearest the under-developed region of the fruit have not developed.

3·3 Gibberellins

3·3·1 *Effects of gibberellins on plants*

The gibberellins show a really spectacular array of effects on plants, many of which are listed in Table 1. Amongst others which may be mentioned are:

(a) the reversal of light-induced inhibition of stem growth
(b) substitution for a cold requirement or a long-day requirement for flowering in many plants
(c) the induction of parthenocarpy, especially in some plants (such as peach and plum) in which auxins do not have this effect
(d) the alteration of sex expression in some plants (e.g. cucumber and a number of mosses) in the direction of maleness. Auxin tends to promote the opposite effect.

3·3·2 *Evidence for the production of gibberellins by plants*

There is abundant evidence that gibberellins are almost ubiquitous in plants, since with the exception of some fungi and some bacteria they have been reported in all groups of plants which have been examined (see WAREING and PHILLIPS, 1970; CLELAND, 1969), though in some lower plants it is not certain whether they have a growth-controlling role.

Many plants have been shown to contain several gibberellins, though there has been little progress in finding out what the various roles of the different substances in one plant may be. Like auxins, gibberellins occur in the 'free' form and also bound to other substances, notably to sugars as glycosides and probably to proteins (see LANG, 1970). They are often present in especially large amounts in young regions such as growing points and young expanding leaves and some studies suggest that this is where they are synthesized. Very large amounts often occur in developing seeds. Some certainly move very freely in the plant (for example they are present in the xylem sap ascending from the roots) but in some cases they appear to be very localized. We do not yet know the reasons for this, and equally we know almost nothing about the movement of gibberellins from one part of the plant to another over short distances.

3·3·3 *Evidence for the hormonal role of gibberellins*

Partly because of the relative ease with which gibberellins can be extracted from plants and their quantity assayed there is a great deal of indirect evidence for their hormonal role from studies which correlate the amount of gibberellin present in the plant with some observable growth phenomenon which is known to be stimulated by exogenous gibberellin. A good example of this is given by studies on flowering in a long-day requiring strain of the henbane (*Hyoscyamus niger*). Plants kept in short days had only a basal rosette of leaves and contained a single gibberellin-like substance. Application of gibberellic acid to the plants rapidly induced stem elongation and flowering. When untreated plants were moved to long-day conditions there was an increase in the amount of gibberellin-like material present, and very clear qualitative changes took place in the types of gibberellin present. Because these changes and the start of stem growth came so close together, it was not possible to be certain whether the observed changes in gibberellins were a cause or a result of the observed stem growth. The same problem besets all studies of this type, though it is clear that the fact that gibberellic acid will substitute for the correct day-length treatment in inducing stem elongation and flowering provides strong supporting evidence for a role for endogenous gibberellins in such processes.

Very similar results can be obtained in experiments with dormant vegetative buds of many plants. Increases in endogenous gibberellins are often associated with the natural breaking of dormancy and exogenous gibberellic acid will frequently cause artificial dormancy breaking.

A particularly useful group of substances which has aided studies on the hormonal role of gibberellins is provided by the synthetic plant growth retardants which have become available in recent years. The formulae of some of these are shown in Fig. 3.5. Biochemical studies on the mode of

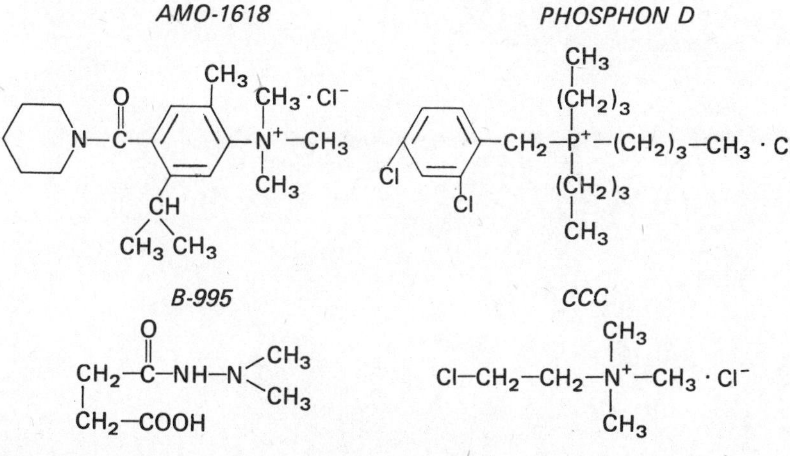

AMO-1618

PHOSPHON D

B-995

CCC

$Cl{-}CH_2{-}CH_2{-}N^+(CH_3)_3 \cdot Cl^-$

Fig. 3-5 Formulae of some synthetic dwarfing compounds. AMO-1618, Phosphon D and CCC are known to inhibit gibberellin biosynthesis in some plants. B-995 has its effect in a different way.

action of these growth retardants have shown that one of the effects of some of them is to block the synthesis of endogenous gibberellins. In some cases the precise step in biosynthesis which is inhibited can be pinpointed. The growth retarding effect of such substances can thus be attributed at least to some extent to the reduction which they cause in the level of endogenous gibberellins in the tissues. This is confirmed by the fact that dwarfism produced by these compounds is often completely reversed by the application of gibberellins. It is of interest that commercial use is made of some such dwarfing compounds in, for example, the production of dwarf chrysanthemums as pot plants with B-995. Caution is needed in the interpretation of results obtained by the use of growth retardants as there is much evidence that they have effects on plants other than those caused by their inhibition of gibberellin biosynthesis, and results can in some cases be highly anomalous. For example, the inhibitor CCC is known to act as a slight promoter of plant growth when applied in low concentration to some plants, and there are recent reports that, contrary to expectations, it actually increases the amount of gibberellin-like substances extractable from pea and tomato plants in spite of its dwarfing effect. The retardant B-995 is not in fact thought to operate directly through an action on gibberellin synthesis, though many of its effects are similar to those of other retardants which do act in this way.

The hormonal control of growth II. Cytokinins, abscisic acid, ethylene 4

4.1 Cytokinins

4.1.1 *Effects of cytokinins on plants*

Many of these effects are listed in Table 1, but amongst others are:

(a) induction of parthenocarpy in some fruits
(b) promotion of cell division in some micro-organisms
(c) promotion of bud formation in leaf cuttings and in some mosses
(d) stimulation of water loss by transpiration in some plants
(e) stimulation of tuber formation in potato
(f) stimulation of growth in some species of algae

4.1.2 *Evidence for the occurrence of cytokinins in plants*

Haberlandt in the early part of the century was the first to produce evidence of a substance in plants which would stimulate cell division, but there is now abundant evidence from work on the growth of plant tissue cultured on closely defined synthetic media that many plant extracts contain cell-division-stimulating factors. A large number of such sources, amongst which are a number of fruits, is listed by FOX (1969); other good reviews are those of HELGESON (1968), LETHAM (1969) and SKOOG and ARMSTRONG (1970). Although most of the substances accepted by physiologists as cytokinins are N^6-substituted adenine molecules there is a variety of other substances which have some cytokinin-like activity in bioassays, sometimes interacting with other factors. An example is diphenyl urea (Fig. 4.1) which, it has been suggested, could have its effect by being built into a cytokinin molecule of the substituted adenine type.

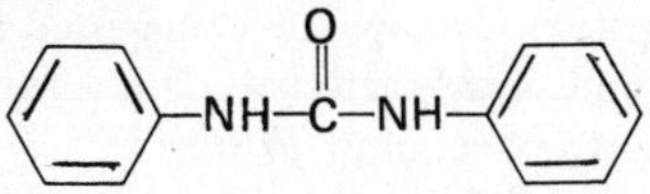

$$C_6H_5{-}NH{-}\overset{O}{\overset{\|}{C}}{-}NH{-}C_6H_5$$

Fig. 4·1 Formula of diphenyl urea.

Table 2 lists some cytokinins which have been isolated and chemically characterized in plants. There seems little doubt that many more will be discovered, though some, such as zeatin, are not specific to one plant and may turn out to be widespread in occurrence.

Table 2 Some naturally occurring cytokinins. For a more comprehensive list which includes chemical names, structures and references, see SKOOG and ARMSTRONG (1970).

Common name	*Source*
Zeatin	Maize grains and other sources
Ribosylzeatin	Maize grains. Fungus culture filtrates. Chicory roots
Dihydrozeatin	Lupin seeds
2iP	*Corynebacterium fascians*
	Various transfer RNA's
6-methylaminopurine	RNA's of various plants

4.1.3 *Evidence for the hormonal role of cytokinins*

Convincing evidence of the hormonal role of endogenous cytokinins is less easy to produce than it is in the case of the other hormones. One reason for this is probably that the extraction and purification of cytokinins from plant tissues is usually a rather long and complicated procedure. Less work of the kind which shows correlations between endogenous cytokinin level and specific growth phenomena is therefore available.

Although exogenous cytokinins will sometimes cause the breaking of dormancy in buds, only the work of DOMANSKI and KOZLOWSKI (1968) on birch and poplar has so far suggested a role for endogenous cytokinins in dormancy breaking, since they found changes of cytokinin-like substances in bud extracts which paralleled growth in buds emerging from dormancy.

Cytokinins are known to occur in the xylem sap of a number of plants This, combined with some of the known effects of roots on shoot growth, has led to the suggestion that cytokinins needed for the control of growth in the shoot are synthesized in the root and transported upwards in the transpiration stream. In 1967 Sitton and his colleagues showed that when the content of cytokinins in the xylem sap of sunflowers was estimated at various times it increased during the period of most rapid growth of the plants but fell to a tenth of its maximum value when the plants had reached their full size (see Fig. 4.2). The workers suggested that this fall in endogenous cytokinins was one of the factors leading to senescence of the leaves. In view of the well known effect of cytokinins in delaying senescence some such role for endogenous cytokinins would seem quite possible.

The effect of cytokinins in promoting lateral bud growth has been noted earlier (Table 1). The bacterium *Corynebacterium fascians* causes the so-called Witches' Broom disease of a number of plants, in which a major symptom is the abnormal outgrowth of lateral buds. The symptoms can be simulated by cytokinin treatment of the plants, and the involvement of endogenous cytokinins is strongly suggested by the fact that the bacterium is known to produce a cytokinin (see Table 2).

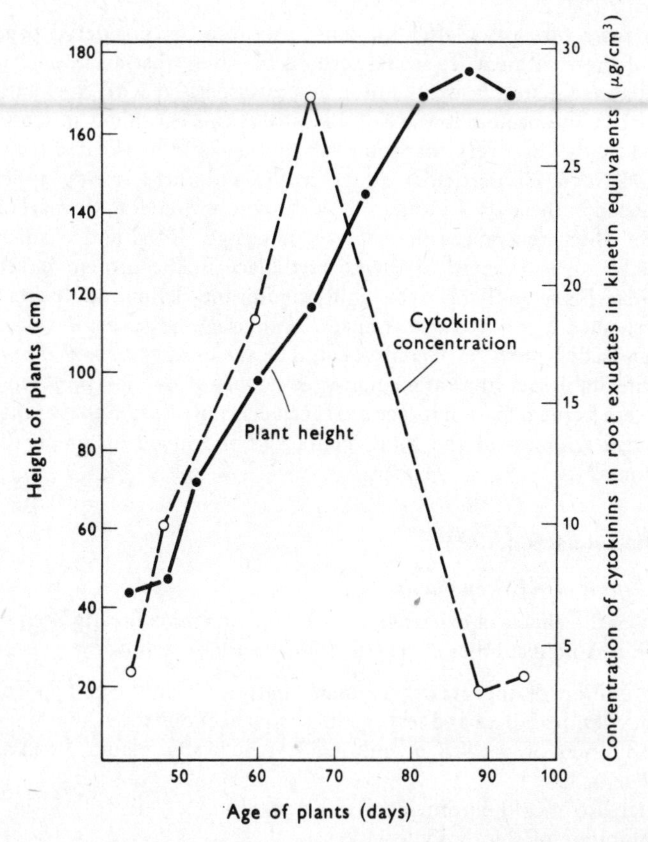

Fig. 4-2 Growth of sunflower plants and parallel changes in endogenous cytokinins in the bleeding sap obtained by cutting off the tops of the plants and collecting the root exudate for 24 hours. (Data of SITTON *et al.*, 1967.)

4.1.4 *Mode of action*

The way in which the control of cell division in plants is achieved is one of the major problems of developmental botany. There has consequently been enormous interest in studies on the precise way in which cytokinins act within the cell ever since their cell-division-inducing properties were first recognized. It is not impossible that these substances may have different modes of action in the different processes which they appear to control. However, there is always a fond hope at the back of the mind of developmental physiologists that they may find cases of genuine 'master reactions' which in one step account for a variety of different effects.

In the case of the cytokinins such hopes have not so far been realized. Many biochemical changes can be shown to follow cytokinin treatment, but most of these take place after too long a time to be considered primary results of the treatment. It seems certain, however, that an important, if not fundamental step in cytokinin action is associated with the synthesis of new RNA and protein in the cell. Because of this enormous interest was attracted by the discovery that some cytokinins could be isolated from the transfer RNA's for particular amino acids, and furthermore, appeared to be located in these tRNA's at a site next to the anti-codon, the part of the molecule which recognizes the code in messenger RNA and enables the amino acid to be inserted in the correct place in the protein molecule. *Prima facie* this seems likely to be highly significant. Serious doubts remain, however, since it is extremely difficult to envisage how a wide variety of effects and a delicate level of control could be achieved by a molecule which would inevitably act in a rather non-specific way if its incorporation into tRNA were a crucial part of its action. It can certainly be said that at present we are still ignorant of the fundamentals of the mode of action of the cytokinins.

4.2 Abscisic acid (ABA)

4.2.1 *Effects of ABA on plants*

Many of the effects of exogenous ABA on plants have already been listed in Table 1. Amongst other effects it may have are the following:

(a) inhibition of the growth of many parts of plants e.g. hypocotyls, radicles, leaf discs and leaf sections, root sections
(b) inhibition of growth in cultures of the water plants *Lemna* and *Wolffia*
(c) increase of cold hardiness in *Acer negundo*
(d) inhibition of chlorophyll synthesis
(e) promotion of germination in spores of some fungi
(f) induction of stomatal closure in some plants.

ABA is remarkably non-toxic except at extremely high, non-physiological concentrations and it is clear that plants can dispose of it by a variety of chemical means, notably perhaps by the formation of an abscisyl glucoside in which the molecule is linked to a glucose residue.

4.2.2 *Evidence for the occurrence of ABA in plants*

ABA has been isolated in a crystalline form from a few sources in higher plants (Table 3). As can be seen from this Table however, the quantities present are very small, and very large amounts of tissue are needed for the extraction of crystallizable amounts. However, studies using GLC and the measurement of the Optical Rotatory Dispersion of solutions of ABA have given fairly certain identifications of the substance in a wide variety of plants. Similarly there is presumptive evidence for its occurrence from detailed chromatographic studies in which plant extracts have been com-

Table 3 Sources and quantities of abscisic acid crystallized and chemically identified in plants. (Data summarized by ADDICOTT and LYON 1969.)

Source	*Amount isolated* (mg)	*Concentration in tissues* (μg/kg)
Cotton fruits	9.0	40
Lupin fruits	5.2	25
Pea pods	9.0	7
Sycamore leaves	0.26	9
Yam tubers	5.5	16

pared with authentic ABA in a variety of solvents and with several bioassays. Table 4 shows some of the plants and plant parts in which ABA has been identified in these ways.

Naturally occurring ABA is in fact one of a pair of enantiomers which have the property of rotating the plane of polarized light in opposite directions. These are known as (+)-ABA (naturally occurring) and (−)-ABA. Synthetic ABA is a racemic mixture of the two enantiomers. The + and − forms have equal activity in at least one bioassay though their metabolic fates in the plant are slightly different (MILBORROW, 1970).

In addition to this optical isomerism ABA also shows geometrical isomerism, and under the influence of ultra-violet light a related molecule,

Table 4 Some sources from which abscisic acid has been identified by physico-chemical methods other than crystallization and subsequent analysis. Most of this work, though not all, is due to MILBORROW (1967).

Plant	*Source of extract*	*Plant*	*Source of extract*
Apple	Fruit	Olive	Fruit
Ash	Leaves, buds	Peach	Leaves
Avocado	Fruit, seed	Plum	Twigs
Birch	Leaves	Potato	Tuber
Bracken	Rhizome	Rose	True fruits, pseudocarp
Cabbage	Heart leaves	Strawberry	Leaves
Coconut	Liquid endosperm	Sweet chestnut	Leaves
Couch Grass	Rhizome	Sycamore	Many parts, including roots
French Bean	Shoot	Willow	Xylem sap
Lemon	Fruit		
Linden	Fruit, seed		
Maize	Fruits		

2-*trans*-ABA is formed (Fig. 4.3). 2-*trans*-ABA is considerably less active than ABA in some systems, though many of its properties are similar, and it is certainly formed in the course of the extraction of ABA from plants in even diffuse sunlight. The proportion of ABA undergoing this rearrangement can be detected by GLC but not by biological means, which introduces additional uncertainties into assays of endogenous ABA in plants. 2-*trans*-ABA has been reported to occur in plants, but only in parts which had been exposed to light before extraction.

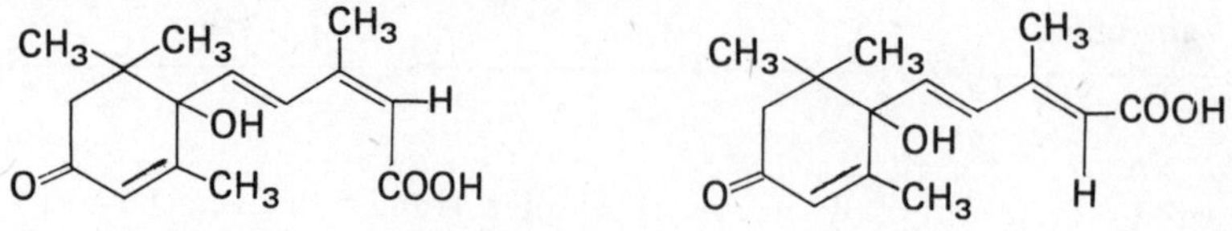

Fig. 4-3 Formula of ABA (left) and 2-*trans*-ABA (right).

4.2.3 *Evidence for the hormonal role of ABA*

There is rapidly growing evidence for the role of ABA as a hormone (see for example WAREING and RYBACK, 1970; ADDICOTT and LYON, 1969) and just two or three examples will be used to illustrate this.

In the early studies of abscission in cotton fruits by the team working in California ABA was extracted from cotton fruits and its level appeared to

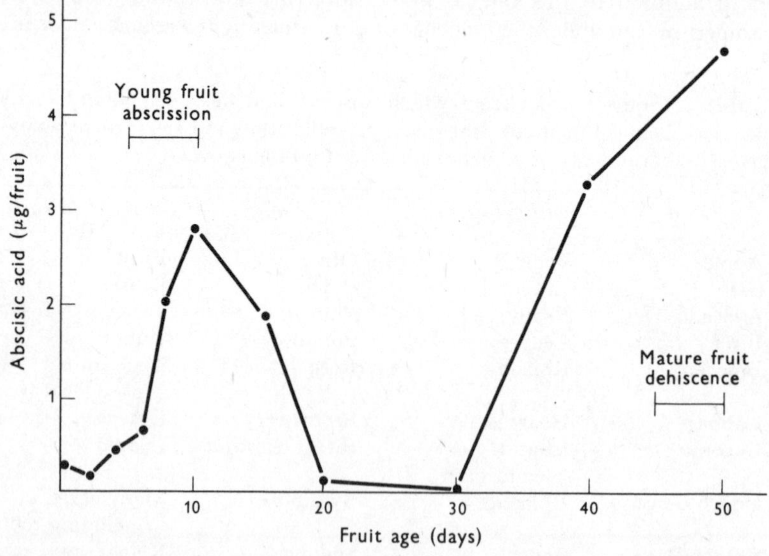

Fig. 4-4 Changes in ABA determined by gas-liquid chromatography during the development of cotton fruits. ABA level rises at the times of young fruit abscission and of senescence and dehiscence of mature fruit. (From ADDICOTT and LYON, 1969.)

increase at times when premature abscission of young fruits or senescence and dehiscence of the older fruits was taking place. The results of a gas chromatographic analysis of this phenomenon are given in Fig. 4.4, which clearly associates both stages of fruit development with rises in endogenous ABA. Application of ABA causes abscission in cotton, and indeed this property was used as a bioassay for the substance. It is clear that this response to ABA is at a site removed from the place of application, and an important requirement for hormonal activity is thereby met.

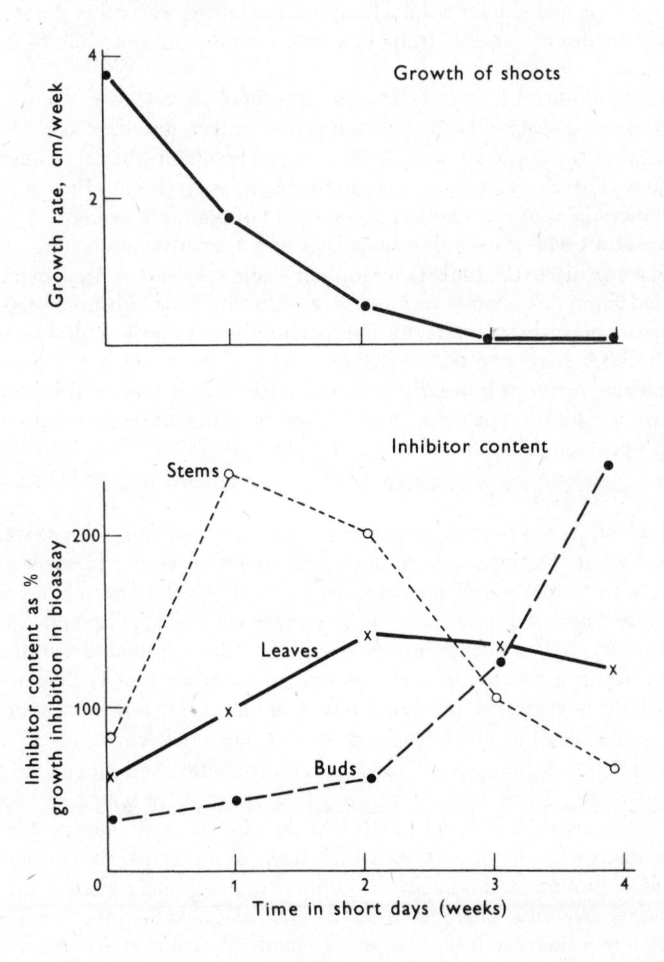

Fig. 4-5 The effect of an increasing number of short days on growth and total inhibitor content of shoots of *Betula pubescens*. Inhibitors were assayed by an oat coleoptile section test and their quantities are indicated by the figures for per cent inhibition of this test on the vertical axis. (KAWASE, 1961.)

It is well known that in many deciduous trees the level of growth inhibitors increases in the leaves as the days shorten in the autumn. This increase of inhibitory materials (the major component of which has been shown to be ABA) is associated with decreasing growth rate in the shoot and the eventual formation of dormant buds (see Fig. 4.5). Wareing and his colleagues at Aberystwyth have shown that authentic ABA, and also extracts of leaves of plants which have been kept under short days, will cause formation of dormant buds on the shoots of actively growing seedlings of sycamore, and similar results have been obtained with other plants. This work provides extremely strong evidence for the hormonal role of ABA in dormancy.

Recent studies by WRIGHT and HIRON (1969) have shown a spectacular effect of the wilting of leaves on their ABA content, and have subsequently shown that the application of ABA to leaves results in the rapid closure of stomata. These observations may indicate an involvement of plant growth substances in a hitherto unsuspected aspect of plant physiology.

One additional growth inhibitor deserves a brief mention here. This is the recently discovered substance lunularic acid (VALIO *et al.* 1969) which was isolated from the liverwort *Lunularia cruciata*. This substance has subsequently been shown to be present in a number of species of liverworts, in which ABA may not occur (PRYCE, 1971). There are some structural similarities between lunularic acid and ABA, but it may well be that the the two inhibitors represent, in evolutionary terms, different solutions of a similar problem by different types of plant.

4.2.4 *Mode of action*

Much effort has been expended on studies of possible modes of action of ABA, and in particular there have been many attempts to show that it inhibits the synthesis of specific nucleic acids or the enzymes produced by these. There is ample evidence that enzyme systems of several kinds are affected by ABA. For example, the gibberellin-stimulated synthesis of α-amylase in germinating barley is strongly inhibited, as is that of many hydrolytic enzymes in the same tissue. This effect, however, seems insufficiently specific to be a crucial primary stage in ABA action.

In radish leaf discs, which senesce rapidly under the influence of ABA, studies of the RNA content showed that again there was a rather non-specific decrease in all types of RNA at the same time. In addition to these problems, many of the studies which have been carried out have shown effects on proteins and nucleic acids only after a time-lag sufficient for many secondary chemical changes to have taken place, and consequently most workers feel the search for the precise mode of action of ABA is likely to go on for some time. It is, of course, very possible that a substance with so many effects on plants may have different modes of action in different systems. Ample references to many of these problems are given in the excellent review article by ADDICOTT and LYON (1969).

4·3 Ethylene

4·3·1 *Effects of ethylene on plants*

Many of the effects of ethylene on plants have already been mentioned in Table 1. Amongst other effects it may have are the following:

(a) induction of epinasty in leaves
(b) inhibition of auxin transport within the plant
(c) inhibition of plumular hook opening and apical bud expansion in pea and bean shoots
(d) stimulation of the synthesis of some enzymes (e.g. peroxidases). Also stimulates release of preformed α-amylase in germinating barley grains.

References to details of many ethylene effects on plants may be found in the review by PRATT and GOESCHL (1969) and in OSBORNE (1968).

4·3·2 *Evidence for the occurrence of ethylene in plants*

As mentioned earlier it has been known for many years that plants can produce ethylene, but most of the earlier work was severely hampered by the difficulty of measuring the very small quantities involved. The method most often used was a bioassay involving the so-called 'triple response' of young pea seedlings, which in response to exogenous ethylene show a reduction of stem growth, a loss of geotropism and a pronounced radial expansion of the region just below the apex. The introduction of gas chromatography has enabled us not only to measure very small quantities of ethylene very accurately, but also to study such things as the time-course of ethylene production by plants or parts of plants. It has even been possible to estimate quantities of ethylene inside plant organs and to show how ethylene production varies from one part of a plant to another. For example, young leaves of the cotton plant produce much more ethylene than the older leaves.

Curiously, one of the organs which does seem to produce a lot of ethylene is the flower. After pollination the flowers of Vanda orchids produce a large amount of ethylene; the trigger which stimulates this is pollination, but a supply of exogenous ethylene will cause the same burst of endogenous production, and once the process has begun it continues naturally and is associated with the fading of the petals. Fruits also produce ethylene, but in this case the quantity is generally small at first and becomes greater as the fruits age.

Ethylene production is frequently stimulated by auxins, both natural and synthetic, and it has also been shown that various types of stress, such as wounding, exposure to ionizing radiation, disease or physical restriction will cause increased ethylene production.

4.3.3 *Evidence for the hormonal role of ethylene*

Although it is a gas at normal temperatures it is not easy to show unequivocally that ethylene moves from the site of synthesis to the site of action. Evidence that the substance does move in the required way was proposed by OSBORNE (1968) who treated leaves with a drop of the n-butyl ester of 2,4-D, which is an effective defoliant. The eventual response to this treatment (abscission of the leaf) was at a site removed from the treatment, and strong evidence was given to suggest that the factor actually causing the response was ethylene moving from the treated area of the leaf (see Fig. 4.6).

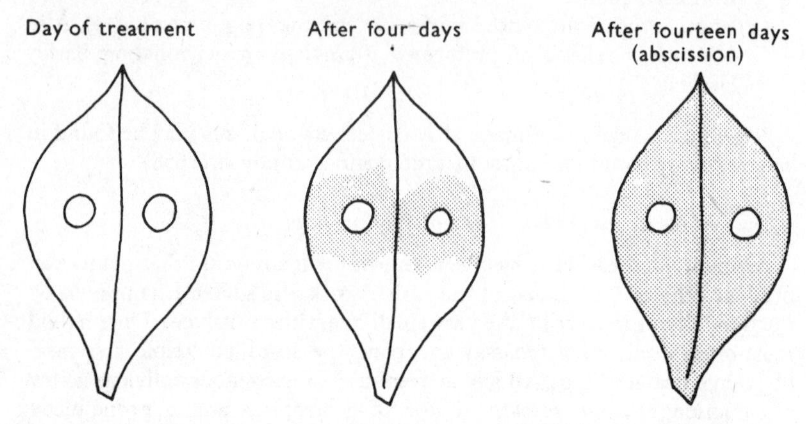

Fig. 4-6 Development of senescence in leaves of *Euonymus japonica* following treatment of the circled areas with drops of n-butyl 2,4-D, which causes rapid evolution of ethylene. Senescent areas are shown shaded, while green areas are unshaded. (After OSBORNE, 1968.)

Evidence for the role of endogenous ethylene in triggering the climacteric rise in respiration (and the subsequent changes in colour, hardness and sweetness) in fruits is very clear. Table 5 shows that in some fruits the level present before the climacteric is as low as or lower than the exogenous concentration needed to cause the climacteric rise (where this takes place at all). At the time of the climacteric the endogenous level is much higher, and it can be shown that this rise in ethylene concentration precedes the onset of the climacteric. Of special interest in the Table are the figures for the two citrus fruits (oranges and lemons) which do not show a climacteric rise unless treated with exogenous ethylene. (See also MAPSON, 1970.)

Ethylene production in geotropically or phototropically stimulated stems appears to be related to the auxin contents of these tissues, but not to be directly responsible for the observed growth responses. In roots, however, there is quite strong evidence (CHADWICK and BURG, 1967; 1970) that ethylene is causally involved in the geotropic response. Geotropic

Table 5 Concentrations of ethylene present in some fruits before and at the time of the climacteric respiratory rise and the threshold concentrations of exogenous ethylene needed for induction of the climacteric. (All concentrations in ppm. After BURG and BURG, 1965)

Fruit	*Prior to climacteric*	*At onset of climacteric*	*Threshold for induction of climacteric*
Avocado (var. Choquette)	0.04	0.5–1.0	—
(var. Fuente)	—	—	0.1
Banana (var. Gros Michel)	0.1	1.5	0.1–1.0
(var. Lacatan)	0.2	—	0.5
(var. Silk Fig)	0.2	0.9	0.2–0.25
Cantaloupe (var. P.M.R. No. 45)	0.04	0.3	0.1–1.0
Honeydew melon	0.04	3.0	0.3–1.0
Tomato (var. VC-243-20)	0.08	0.8	—
Orange (var. Valencia)	0.1	—	0.1
Lemon (var. Fort Myers)	0.1	—	0.1

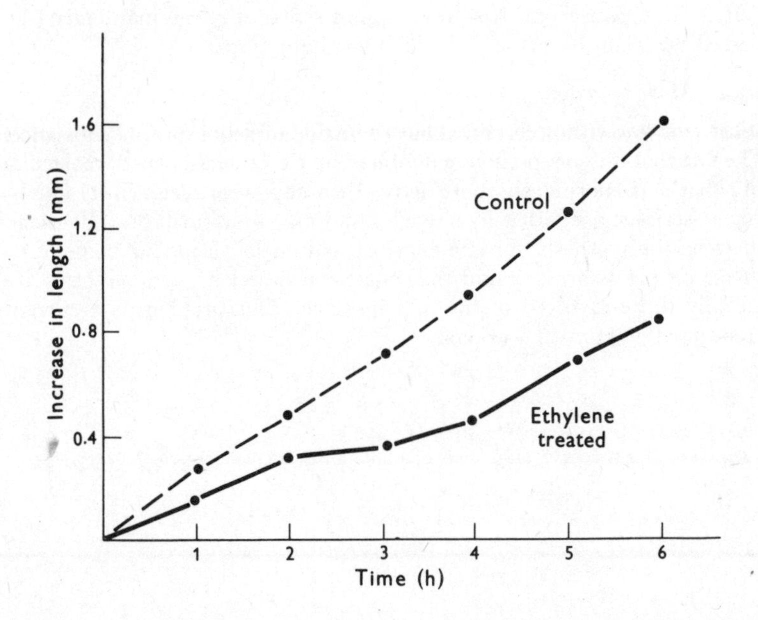

Fig. 4-7 Diagram showing the effect of high concentrations of ethylene on the growth of pea root segments. Ethylene was applied at the start of the experiment. A clear difference between the control and the treated segments is apparent after only one hour. (After CHADWICK and BURG, 1967.)

stimulation causes a rapid rise in ethylene production, and ethylene itself has an almost instantaneous effect on root growth (see Fig. 4·7). More significant, however, in the interpretation of the role of ethylene here is the fact that CO_2, known to be a competitive inhibitor of ethylene in many processes, will prevent auxin-induced root growth inhibition. All this evidence strongly suggests that in roots geotropic stimulation causes auxin redistribution; enhanced auxin on the lower side of the root causes ethylene production and the ethylene inhibits growth on the lower side of the root, causing downward curvature. When roots have become vertical again ethylene production returns to normal and any remaining excess can rapidly diffuse away.

As a postscript to this section it is fascinating to consider the experiments of GOESCHL *et al.* (1966) with pea seedlings. When seedlings were grown in tubes so that they were physically constricted ethylene production was greatly increased. Since the ethylene could not diffuse away, this caused the typical response of radial expansion of the shoots. Since a thick shoot can presumably exert a greater vertical pressure than a thin one this suggests a method by which the plant may be enabled to withstand or overcome physical stresses such as those caused by bursting through the soil crust on germination.

It is clear even from these few examples that ethylene may indeed be classed as a plant hormone alongside the other groups.

4·3·4 *Mode of action*

Little is known for certain about the mode of action of this substance. The fact that it is competitively inhibited by CO_2 (another small molecule) and that it is enormously more active than any other olefin in its homologous series suggest that its site of action may be a small one. It has an effect on the synthesis of some enzymes and nucleic acids but some of its effects on plants are so rapid that nucleic acid and protein synthesis are unlikely to be involved in the first instance. This problem is a current preoccupation of many workers.

5 Interactions, multiple actions and sequential actions of plant growth substances

Up to now we have considered the effects and internal roles of plant growth substances on a substance by substance basis as far as possible, and while processes in which more than one substance is involved have been mentioned they have been passed over rather rapidly. The more we get to know about the hormonal control of growth, however, the more it becomes clear that it may be very misleading to consider one class of substances at a time. Table 1 shows that many processes are affected by more than one class of exogenous substance, and there is every reason to suppose that many processes may be internally controlled by two or more substances acting together in some way.

Before considering some examples of the evidence bearing on this point it will be useful to look briefly at some of the ways in which two substances might affect any one process. This is important since there is some confusion in the literature about the terminology of such effects, and especially about the term 'interaction', which is used in two quite distinct senses.

Let us consider two substances, A and B, which are thought to be involved in the control of stem growth. If these are applied to an appropriate plant separately and together there are several possible results which might be expected. Some of these are set out in Table 6.

Table 6 Five possible results of an imaginary experiment in which two substances thought to affect stem growth were applied independently and together to a suitable test plant.

Treatment	*Mean height (cm) of test plants after a standard period*				
	Case (i)	*Case (ii)*	*Case (iii)*	*Case (iv)*	*Case (v)*
Control	10	10	10	10	10
Substance A	12	12	12	10	10
Substance B	12	6	12	12	10
A + B	14	8	18	18	18
Effect	Additive	Antagonistic (Additive but opposite effects)	Positive interaction	Positive interaction	Positive interaction
Remarks	Effects of A and B could be quite independent			Effects of A and B synergistic	

The greatest terminological problem is given by Case (ii), in which the two substances are having opposing effects. Such situations have frequently been described in the literature as 'interactions', though statisticians would confine this term strictly to Cases (iii), (iv) and (v) in the Table, and it might be better if physiologists were to do the same. The term interaction cannot really be applied to Case (ii) except where the modes of action of both substances are known and are what is called 'competitive'. In such cases the antagonism between substance A and substance B would be due to the fact that they were in some way competing for the same site of action. It is one of the characteristics of such cases that a saturating dose of one substance will prevent the response to the other, and in such situations the term 'competitive interaction' may be used. In all other cases it would be much better to stick to the term 'antagonistic' or 'additive but opposite' to describe the relationship.

It follows that merely to say that two substances are antagonistic in a given system does not tell us anything about the reasons for the antagonism. In the same way, to say that two substances interact (either positively or negatively) does not tell us anything about the reason for the interaction. In Case (v) from Table 6 it might be that the substances A and B need to combine together in some way to produce the response. In this case neither could operate without the other. Alternatively one may have an effect on the plant which does not in itself lead to an increase in the rate of growth, but which is necessary for the other to be able to work to produce this increase.

Case (i) may also be explained in more than one way. The substances A and B may be working quite independently on 'unrelated' aspects of growth at the same time, or they may be affecting different stages of the growth process to produce what is essentially a sequential response. Only further experiments would distinguish between these possibilities.

From what has been said so far it is clear that the greatest caution is needed in interpreting and discussing the results of experiments in which two substances are being considered together. If three substances are likely to be involved it is clear that the problem becomes even more complicated. In this book we have been dealing with five classes of substance, and some growth processes are affected by all of these. Need one say more?

Both the interest and the difficulties of studies on processes involving more than one substance at a time are perhaps best illustrated by considering some examples of experiments of this kind.

5.1 Effects of ABA and a cytokinin

A spectacular example of an antagonistic relationship between two substances is given by work on the water plant *Lemna*. When this plant was grown in a culture medium under controlled conditions growth could be virtually stopped by adding ABA to the medium; growth was resumed on the addition of appropriate concentrations of the synthetic cytokinin benzyl adenine. By suitable manipulation of concentrations of these two substances the growth of the plant could be effectively turned on and off at will. Such studies suggest, but do not, of course, prove, that the growth of *Lemna* may be regulated naturally by a balance of endogenous cytokinin and inhibitor concentrations.

5·2 Effects of ABA and gibberellins

The other obvious example of an antagonistic relationship between two substances is that between ABA and gibberellins in the control of dormancy in plants, ABA tending to increase in amount as dormancy is induced, and to decrease as it is broken, whilst gibberellins show the opposite behaviour. There are other processes in which ABA and gibberellins appear to have the same sort of antagonistic relationship, but dormancy has been one of the most studied. In fact it seems very probable that natural dormancy is regulated by the internal balance of ABA (and possibly other inhibitors) and gibberellins, the former exerting the main influence at the onset of dormancy and the latter at its end. There is evidence that it is an increase in endogenous gibberellins rather than a decrease in endogenous inhibitor levels which brings about dormancy breaking in buds; at the onset of dormancy levels of both substances tend to be changing simultaneously in the expected directions.

5·3 Interactions between IAA, GA_3 and kinetin (K) in stem growth and the movement of metabolites

When we come to consider evidence for actual interaction between plant growth substances there are many possible examples upon which to draw. One of these, which shows two different kinds of interaction, is illustrated in Fig. 5.1. This involves the responses of isolated segments from the young shoots of light-grown pea seedlings to IAA, GA_3, and K. Segments were floated on the solutions and at appropriate concentrations both IAA and K caused an increase in longitudinal growth in 24 hours, while GA_3 did not. However, IAA and GA_3 interacted to increase the length of the segments (positive interaction similar to Case (iii) in Table 6 above). A different type of interaction occurred between IAA and K, since the length increase promoted by the two together was less than that caused by IAA alone, but positive interaction did take place in the promotion of growth in diameter, and thus in volume, of the segments. This illustrates that interaction is not confined to the amount of growth, but may also influence its direction.

It is known that the presence of developing fruits on a plant has a considerable effect on the movement of other substances within the plant; there is substantial movement of photosynthates from the leaves to the fruit, for example, and a similar movement of many minerals. SETH and WAREING (1967) used French Bean plants to show that some of this effect of the fruit

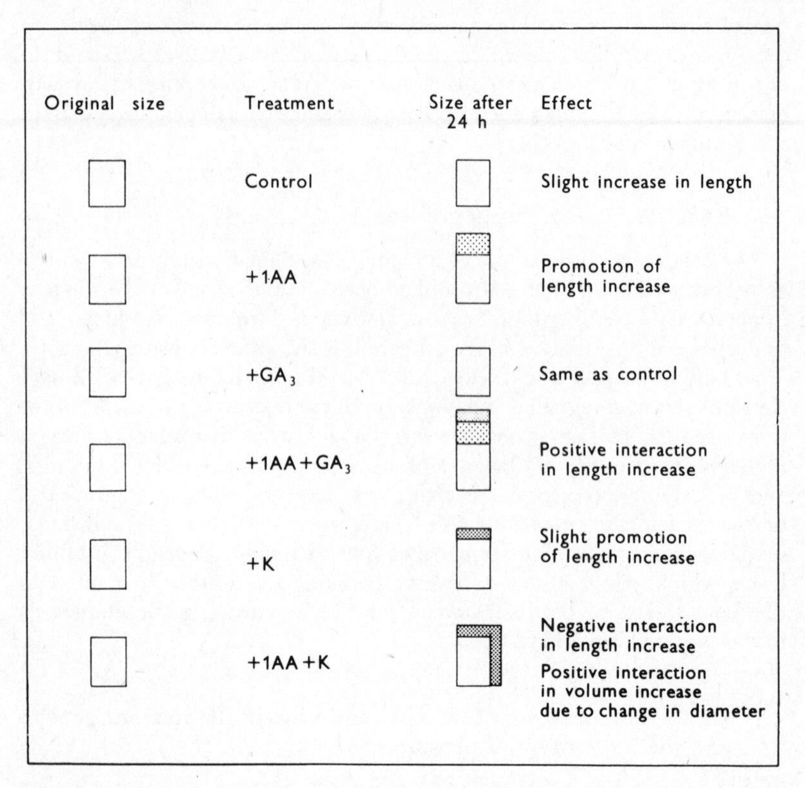

Fig. 5-1 Interaction between IAA, GA_3 and kinetin (K) in promoting the growth of 1 cm segments cut from green pea seedlings and floated on solutions of the growth substances for 24 hours. (After OSBORNE, 1965.)

in drawing metabolites from other parts of the plant might be due to the influence of hormones.

Developing fruits were cut from the plants and the hormones applied to the cut ends of the peduncles. Movement of radioactive phosphorus (^{32}P) applied lower down the stem was analysed by measuring the radioactivity which accumulated in the peduncles. IAA interacted strongly and positively with both K and GA_3 in causing an increase in movement of ^{32}P into treated peduncles. Similar experiments (Table 7) using decapitated plants and measuring movement of ^{32}P into the decapitated internode showed the same effect, and in this case there was an even larger effect when all three substances were applied together. In view of the fact that both young fruits and stem apices are rich sources of plant growth substances it seems very probable that interactions of this type do have a significant part to play in the endogenous control of the movement of substances from one part of the plant to another.

Table 7 Effect of IAA, GA_3 and K on the accumulation of radioactive phosphorus in the decapitated internodes of bean plants. (SETH and WAREING, 1967)

Treatment	*Radioactivity as % of Control*
GA_3	358
K	246
IAA	1921
IAA + GA_3	3600
IAA + K	3768
IAA + GA_3 + K	8233

5·4 Interaction between GA_3 and K in controlling lateral bud growth

An example of interaction which illustrates a different point is given by some experiments on apical dominance in the tomato plant (CATALANO and HILL, 1969). The presence of the apical bud in many plants prevents the outgrowth of lateral shoots. Removal of the apex releases this inhibition, and axillary shoots may grow out. Replacement of the excised apex by a high concentration of IAA frequently prevents such outgrowth, the IAA thus substituting for the apex in its control of this process (Fig. 5.2). Direct treatment of lateral buds of some plants with kinetin or other cytokinins has been shown to cause lateral buds to grow out even in the presence of the apex or of an exogenous source of IAA in its place. In the tomato plant this was not found to be the case, for although there was sometimes a small effect of K on lateral bud growth this was enormously enhanced by applying K and GA_3 together (Fig. 5.3). This was found largely with decapitated plants treated at the cut apex with IAA, but the same response also occurs in intact plants.

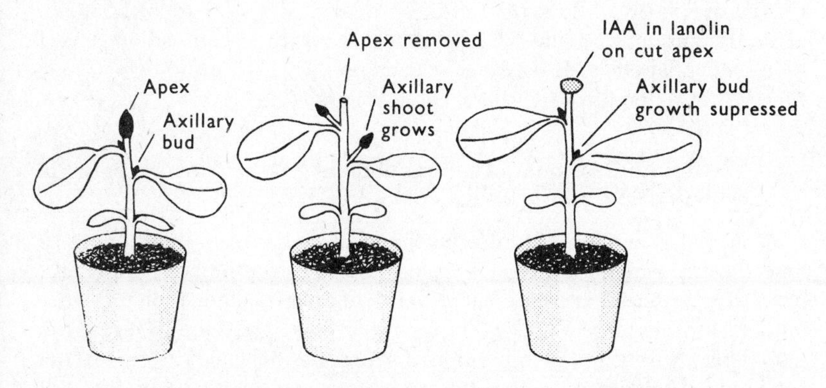

Fig. 5-2 Diagram showing the substitution of IAA for the apex in controlling the growth of lateral buds.

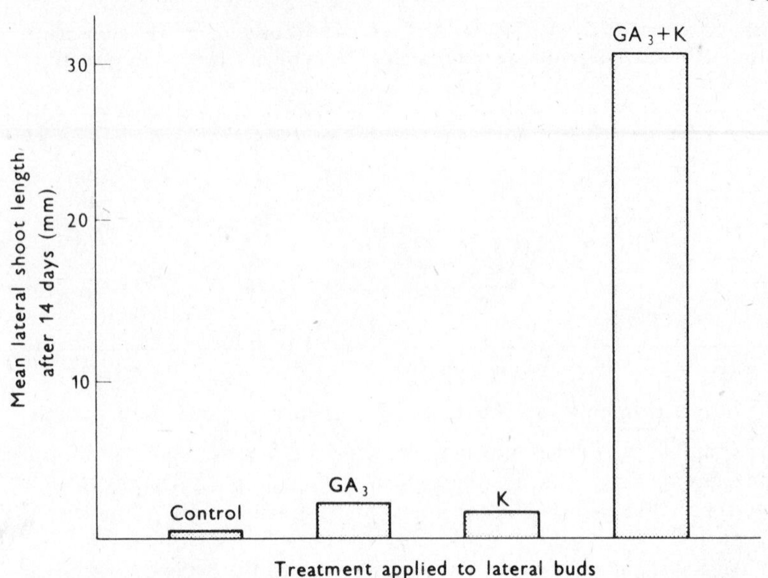

Fig. 5-3 Interaction between GA_3 and K in promoting lateral shoot growth in tomato seedlings decapitated and treated with IAA at the cut apex. (Data of CATALANO and HILL, 1969.)

In an attempt to analyse the phenomenon a little further plants were decapitated, treated with IAA on the cut surface and with either K or GA_3 on the lateral buds. After various intervals the growth substance solutions in lanolin were carefully removed from the lateral buds and replaced by a mixture of K and GA_3. The results of this experiment were that more rapid outgrowth of lateral shoots was subsequently seen in plants which had been treated previously with K than in those treated with GA_3 or with lanolin alone. It thus appears that here we are dealing with an interaction caused by a sequential effect, K causing a change in the axillary bud, which may then be acted upon by GA_3 to accelerate its growth.

5·5 Multiple and sequential actions of plant growth substances in the growth of the young wheat coleoptile

As a final example of studies involving more than one hormone at a time the experiments of Wright on the wheat coleoptile are particularly interesting, since they represent a very thorough examination of what initially appears to be a rather simple system (WRIGHT, 1968). During the period between the initial imbibition of water and the end of its growth the coleoptile of the germinating wheat grain grows substantially by both cell division and cell expansion. Growth is initially due to cell expansion; this is followed by a period during which cell division occurs as well and finally a period in which growth is again mainly due to expansion of previously formed cells. It was postulated that different growth substances might be involved at different times in this sequence, and this was tested by treating coleoptiles excised at different ages with optimal concentrations of IAA, GA_3 and K and observing their subsequent growth. Figure 5.4 shows the results obtained. The coleoptiles were most sensitive to GA_3 18 hours after the start of imbibition (the earliest time at which they could be excised). Their greatest sensitivity to K was after 30 hours, just prior to the period of maximum cell division, and to IAA after 54 hours, during the final phase of cell expansion. It is during this last phase that coleoptiles of wheat and oats are used as test organs for the bioassay of IAA.

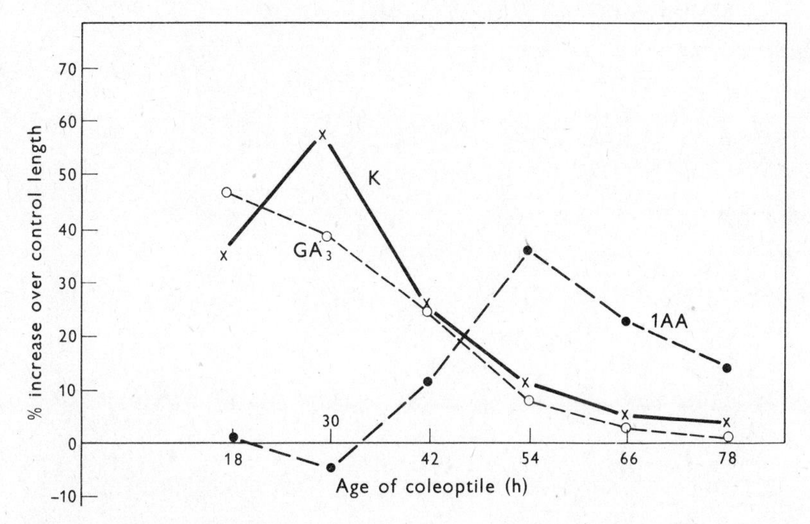

Fig. 5-4 The response of wheat coleoptiles of different ages to IAA, GA_3 and K, showing that the optimal response to each substance occurs at a different age. (Data of WRIGHT, 1961.)

Wright has interpreted these results as suggesting that the growth of the young coleoptile may be maintained endogenously by a continuously changing pattern of growth substances in which gibberellins predominate in the early stages, cytokinin in the middle and IAA in the later stages. In the initial experiments only one exogenous substance was applied at a time and the possibility of interactions between the substances was not studied, though it was of course possible that in the early stages, for example, exogenous GA_3 might be interacting with endogenous IAA or other substances.

In later work Wright studied the growth of young coleoptiles excised 24 hours after the start of imbibition, during the phase when GA_3 and K are active but IAA is not. In these experiments high concentrations of

IAA reduced the effect of both GA_3 and K when applied with them (negative interaction) but the effects of GA_3 and K were shown to be additive. This suggests that if the endogenous IAA in the coleoptile at this stage is at a high enough level it may well modify the effects of endogenous gibberellins and cytokinins by interaction, though the effects of the last two growth substances might be due simply to progressive changes in their ratio, their effects remaining additive.

The experimental difficulties of studies on the endogenous control of growth by substances which may be acting together in one of the ways described are very great indeed, and this is one of the areas of plant growth substance research in which much remains to be done and which presents great challenges to those research workers involved.

A digression: some experiments with plant growth substances — 6

In the past both the high cost of some of the chemicals and the difficulty of handling the very small quantities involved, together with the problems of contamination of glassware and plant material have deterred many people from doing simple experiments with plant hormones. As far as cost is concerned, IAA and several auxin-like growth regulators such as indole butyric acid (IBA) and 2,4-D, GA_3 and kinetin are now all reasonably inexpensive, especially as such small quantities are needed. Other cytokinins and abscisic acid are still rather expensive and quantitative experiments with ethylene are largely impractical because of the analytical equipment needed.

Table 8 The handling of stock solutions of plant growth substances and growth regulators.

Substance	*Solvent*	*Problems*	*Comments*
IAA, IBA	Water	Dissolve with difficulty. IAA is broken down by light—store in dark, preferably in refrigerator.	Warm, or dissolve in minimum quantity of dil. KOH to make K salt, or first dissolve in minimum quantity of alcohol and then dilute to required volume with water.
	Lanolin	Solvent very sticky; takes time to dissolve.	Warm and stir very well. Preferably do not use same day.
GA_3	Water	Dissolves with some difficulty.	Warm and shake well or dissolve first in alcohol as for IAA above. Fairly stable at low temperatures except at low pH.
	Alcohol	—	High concentrations can be prepared and thus useful for application of very small volumes to plants.
	Lanolin	As for IAA above.	As for IAA above.
Kinetin	Water	Very sparingly soluble. Stock aqueous solutions of >50 ppm are not practicable.	Use warm water or dissolve first in alcohol as for IAA and GA_3 above. Stable, especially at low temperatures.
	Lanolin	As for IAA and GA_3 above.	As for IAA and GA_3 above.

Some simple experiments which are possible with plant growth substances, and which use a minimum of special facilities, are described in outline below. Almost all of these form useful starting points for open-ended investigations since the results almost always suggest other practicable experiments which could be done. It is important, however, not to be too concerned if occasional experiments fail or give highly anomalous results. This happens to everyone and we can often learn more from finding out what went wrong than from merely noting an expected result.

A vast amount of relevant practical information is given by MITCHELL and LIVINGSTONE (1968), and Table 8 outlines some details of the handling of certain plant hormones and growth regulators. Stock solutions are best made up at relatively high concentrations (e.g. 100–1000 ppm where possible) and diluted for use. Many hormonal effects occur on treatment of individual plants with amounts of hormone between 0.01 and 10 μg or more per plant.

Small excised parts of plants must normally be floated on aqueous hormone solutions. Seeds can be treated by sowing on filter paper moistened with the solution. In both these cases precautions against evaporation of water are important.

Intact plants can be treated directly with small drops (0.05–0.01 cm^3) of aqueous solutions, preferably including a wetting agent. If very small drops (<0.01 cm^3) are used alcoholic solutions can also be applied direct to plants without causing damage. This technique is useful as high concentrations of hormones are more easily prepared in alcoholic solutions.

6.1 Experiments with auxins

(a) The effects of IAA on cell growth can be seen in a matter of hours if the hormone is applied in lanolin to the leaf axils, or asymmetrically to the stem of a suitable plant (e.g. tomato). Epinasty and stem distortion result, though the effects may be transitory with low doses.

(b) The effect of IAA on cambial division and xylem differentiation can be seen in transverse sections within 1–2 weeks if a short length of the stem of any young dicotyledonous plant (e.g. tomato, bean) is ringed with IAA in lanolin (use 100 ppm or more in initial experiments). Other hormones can be combined with IAA in such experiments.

(c) Many experiments on apical dominance (see for example CATALANO and HILL, 1969) are possible using solutions of hormones in lanolin to substitutes for the apex in decapitated plants or to treat lateral buds directly. It is useful to note that IAA may only replace the apex in stopping lateral bud growth if it is applied at high concentrations (often >100 ppm) and if the plants are grown at low levels of nutrition. This is easily achieved by lowering the nitrogen level in water or sand culture.

(d) Experiments on the effect of IAA and other auxins on the growth of coleoptile and pea stem segments are possible with quite simple equipment, but are more difficult from the manipulative point of view than those already described. Detailed directions on which to base such experiments are given by MITCHELL and LIVINGSTONE (1968).

6.2 Experiments with gibberellic acid

(a) The many effects of GA_3 on whole plants offer endless scope for experiment. Most plants will show visible responses of some kind to GA_3, but one of the best is the readily available dwarf pea cultivar 'Meteor'. Visible growth responses in treated seedlings can be seen in 3–4 days at doses of 0.001–10 or more μg per plant. The larger the dose the longer the induced growth effect continues. Not only stem height but internode number, leaf area, cell division in the sub-apical meristem and many other morphological and anatomical manifestations of growth may be investigated in whole plants treated with gibberellins.

(b) The germination-accelerating effect of GA_3 can be investigated in many plants, but is especially interesting in light-requiring seeds such as those of the lettuce cultivar 'Grand Rapids'. Such seeds only become light-sensitive during imbibition, and may thus be sown on filter paper moistened with water or GA_3 solutions in petri dishes which can be immediately wrapped in aluminium foil to exclude light. Germination is usually visible in 1–2 days and GA_3 concentrations between 1 and 100 ppm should be tried in initial experiments. Dishes can be examined by the use of a green safe-light if they are to be observed more than once (an electric torch covered by several layers of green cellophane is quite satisfactory).

(c) Many interesting experiments are possible on gibberellin-stimulated α-amylase production in cereal grains. Full details of techniques are given by FREELAND (1972) and by COPPAGE and HILL (1973). The best results have so far been obtained with barley. Grains are cut in half, the embryo-containing end discarded and the endosperm end placed cut surface downwards on a dish of starch agar. The enzyme diffuses from the grain and breaks down the starch. Pre-soaking the grains in GA3 or incorporating the hormone in the agar greatly increases the production of α-amylase, and the effect of the enzyme can be made visible either by incorporating iodine in the starch agar or by flooding the agar surface with dilute iodine solution after incubation for one or two days. A clear area around the grains indicates amylase activity and the size of this area gives some measure of the magnitude of the effect. Precautions are needed to maintain sterility in the dishes, and grain samples vary considerably in their response.

6.3 Experiments with kinetin

(a) Kinetin dissolved in lanolin at concentrations of 100–500 ppm may be used, with IAA, in apical dominance experiments as discussed above.

(b) The prevention of chlorophyll degradation in the dark by kinetin can be well shown in some leaves (e.g. Radish, wheat cv. 'Eclipse'). Discs or short (1 cm) segments of leaves are floated on kinetin solutions (0.01–10 ppm) in the dark. Chlorophyll retention after 3–4 days can be scored by eye or by extraction of chlorophyll and spectrophotometry if equipment is available. Many other suitable plants probably exist—it is just a question of trying out new ones.

The mechanism of action of plant growth substances 7

Before plunging into this fascinating but rather complex area of the growth substance physiology of plants we should perhaps consider briefly what is meant by the term 'mechanism of action' in this context. When physiologists study the effect of any substance on any process, what they are always looking for is the very first event in which the substance seems to be involved. For example, it has been known for many years that one of the effects of IAA on coleoptile tissues is to cause a softening of the cell walls. As a result of this the so-called 'wall pressure' of each cell falls. If the osmotic potential of the cells stays the same the cells take up water due to the resulting rise in the water potential depression (sometimes also called suction pressure or diffusion pressure deficit) of the cells. The water uptake causes increase in the volume of the cells and the cells thus 'grow'. Subsequently there is synthesis of new cell wall material. We could say, therefore, that the auxin-induced increase in cell size in this system is due to the effect of auxin on cell wall softening. This in itself would not satisfy a physiologist concerned with the mechanism of action of auxin, who would immediately want to know *how* the auxin has this effect on the cell wall. Is the effect a direct one caused by some sort of attachment of the auxin molecule to the wall, or is wall softening merely the end of a chain of events triggered by the auxin in some different way?

Most of the evidence we have on the mechanism of action of plant growth substances is very fragmentary, and it always relates to specific substances, tissues and conditions. Many of the substances we have considered have a wide range of effects in plants, and it is not by any means certain that the same substance always acts at the same primary site. In a very stimulating account of some aspects of this topic VAN OVERBEEK (1966) has suggested that, just as one key may open several doors, there may indeed be several primary sites of action for a given hormone.

However incomplete our evidence may be we obviously know rather more about the mode of action of some plant growth substances than of others. For this reason this chapter will be concerned almost entirely with some aspects of studies on auxins and gibberellins. In Chapter 4 some reference has been made to the mechanisms of action of cytokinins, ABA and ethylene, and the fact that they are not also dealt with here does not mean that they are not equally important from the point of view of the plant.

Even in the case of auxins and gibberellins it will be necessary to concentrate on what may appear to be some rather small and specialized areas of study and merely to note that any of these may or may not be relevant to other situations.

7.1 The mechanism of action of gibberellins

Some studies on gibberellins will be considered first, partly because almost all the relevant work is fairly recent and partly because of the apparent clarity and relative ease of interpretation of some of its results.

The system which has been most widely used in studying the mechanism of action of gibberellins is the gibberellin-dependent production of the enzyme α-amylase by barley grains. The work of Paleg and his collaborators in Australia (e.g. PALEG, 1965) and of Varner and others in the United States (e.g. CHRISPEELS & VARNER, 1967a, b) is mainly responsible for our detailed knowledge of the barley grain α-amylase system. There has been an element of luck in the fact that this system has turned out to be so useful for the purpose, in that it is one which is relatively easy to handle experimentally and that the response to applied gibberellins (very often but by no means always GA_3) has appeared to be clear, specific and unambiguous, at least until relatively recently.

Let us first consider a few facts about the production of this enzyme in fully imbibed barley grains.

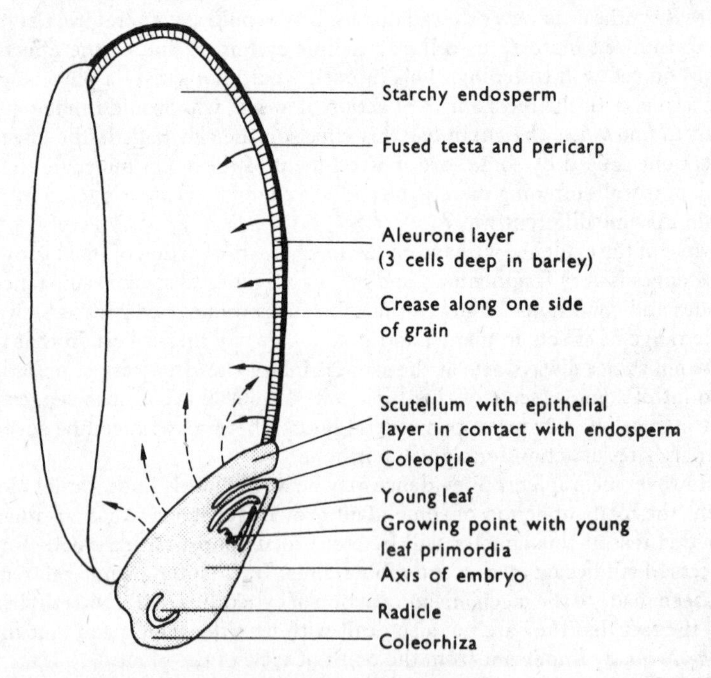

Fig. 7-1 Longitudinal section through a barley grain to show the parts of the embryo, the aleurone layer and the endosperm. Production of gibberellin is indicated by broken arrows, and production of α-amylase in response to this by solid arrows.

(i) The breakdown of starch to sugar in the endosperm is apparently gibberellin-dependent. In grains with the embryo removed the degree of this dependence varies somewhat; in intact grains the necessary gibberellin comes from the developing embryo itself.

(ii) The presence of the aleurone layer is essential in de-embryonated grains, though where the embryo is present some starch breakdown is caused by α-amylase production in the scutellum. (See Fig. 7.1.)

(iii) The gibberellin-induced response is not abolutely specific to α-amylase though most work has concentrated on this enzyme. Other hydrolytic enzymes such as proteases and ribonucleases are also produced. Since it is clear that sugars (and indeed other substances) are needed by the growing embryo during germination, a system in which the embryo partly controls the supply of sugars from the endosperm by its own gibberellin production is a very interesting one in itself.

It was clearly necessary to distinguish in the response whether production of sugars was due to

(i) activation or release of a pre-formed enzyme in the aleurone layer
(ii) the production of a new α-amylase from existing enzyme-specific messenger RNA (mRNA) in these cells or
(iii) the production of new enzyme from new mRNA itself produced under the influence of gibberellin.

Incubation of de-embryonated grains with GA_3 in solutions containing radioactive amino acids led to the isolation of α-amylase which was uniformly radioactively labelled, eliminating possibility (i), since the labelled enzyme had clearly been synthesized during the period of incubation.

To distinguish between (ii) and (iii) it was necessary to compare the effects of inhibitors of protein synthesis on the RNA template (e.g. puromycin and cycloheximide) and an inhibitor of RNA synthesis itself (e.g. Actinomycin D). In situation (ii) this last inhibitor should not affect the gibberellin-induced response, at least for some time. In fact Actinomycin D does largely prevent the response, which therefore appears to be due to the synthesis of new mRNA. Figure 7.2 illustrates these relationships.

The interest of this result rests mainly on the well established view that a single gene in the nucleus probably controls the production of a single type of enzyme-specific mRNA molecule, and hence the production of a single enzyme. It would appear that in this case GA_3 is causing a gene to function when it would otherwise not do so. In the jargon of the cell biologist, GA_3 de-represses the gene, and can thus be said to be acting at a very fundamental level in the cell. Even so, however, we are left with two equally fundamental and unanswered questions. Firstly, *what* exactly happens when the gene is de-repressed and secondly, *how* does the hormone affect this process?

The first question embraces one of the most crucial problems in cell biology and certainly cannot be answered fully, but recent work has given some indications of one of the ways in which control of gene activity may be achieved. It is known that in the chromosomes themselves the DNA is

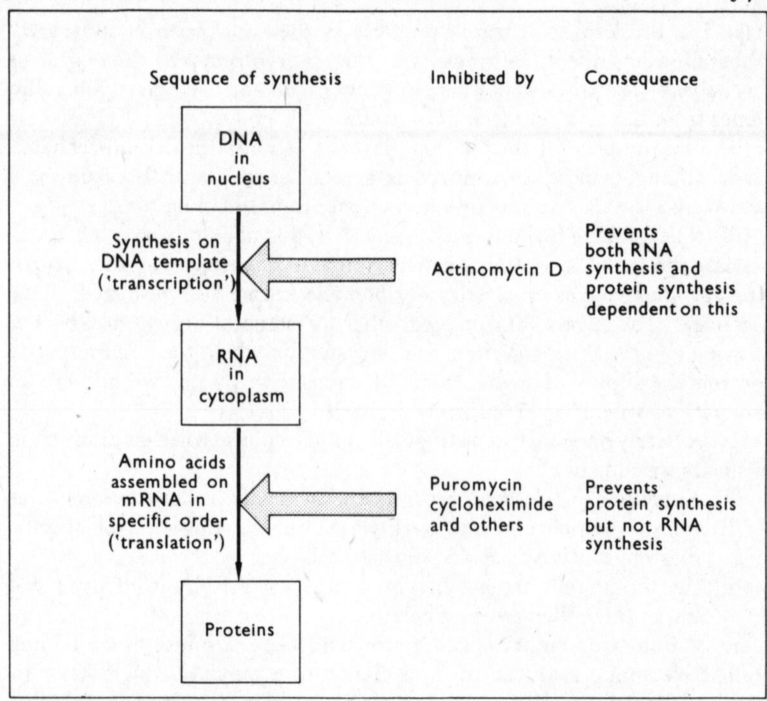

Fig. 7-2 An outline of the principal events of protein synthesis in the cell, showing the points at which the process may be inhibited by Actinomycin D and other inhibitors.

closely wrapped round with protein molecules, the nucleo-histones, which were originally thought to have a largely structural role. It has been suggested that these proteins may in fact physically interfere with the operation of the genes by simply lying close to the DNA, so that when the histone is removed or loosened from its position the DNA is able to begin to function as a template for RNA synthesis. If this is the case in the barley grain system the primary action of GA_3 must be thought of as the mediation of the unwrapping of the appropriate gene for the production of α-amylase-specific mRNA (and indeed of certain other specific genes as well, as already mentioned). Exactly how this might be achieved, our second unanswered question, remains entirely unknown.

There is suggestive evidence from a rather unexpected quarter that hormonal effects may be mediated through a direct effect of the hormone on chromosomal structure and activity. In certain cells of insects such as the fruit fly *Drosophila* and the midge *Chironomus* the chromosomes are extremely large, and individual regions of these, thought to correspond in position to individual genes, can readily be seen. When certain genes are active, for example during metamorphosis of the insect, the relevant regions of these giant chromosomes bulge out to form what are known as 'chromosome puffs' (see BEERMAN and CLEVER, 1964). Puffing can be induced by injection of the insects with appropriate hormones, and it is associated with rapid RNA synthesis at the site of the puffs. It appears that in the puff the strands of DNA spread out and become accessible for RNA synthesis, perhaps as it were slipping out from beneath their histone blanket. In this connection it is very curious, though possibly fortuitous, that GA_3 will induce moulting when injected into locusts in place of the locust moulting hormone ecdysone. Whatever the significance of this observation it is clear that some hormones can apparently act in a way which seems to be directly related to gene structure and function.

As far as the barley α-amylase system is concerned, one could conveniently invoke some such mechanism for GA_3 action if

(i) the kinetics of the response were appropriate (i.e. if the speed of the response and its time lag were adequate to allow for the postulated mechanism to have operated) and

(ii) if the response were specific to GA_3, or at any rate to the gibberellins in general.

In this case the kinetics of the response do not conflict with the postulated mechanism, but recent work in the United States has shown that other substances, especially cyclic 3′,5′-adenosine monophosphate (cyclic 3′,5′-AMP) and certain intermediates in the Krebs cycle, or substances related to these, will also cause α-amylase synthesis in de-embryonated barley grains. Cyclic 3′,5′-AMP is a known mediator of hormone action in some animal systems and it may turn out to have a similar role in plants. These effects were noted only at higher concentrations than those needed for hormones such as GA_3, but it is clear that much further work is needed before we can be categorical about the mode of action of gibberellins even in this system which, it will be recalled, we started off by describing as having given results which were relatively clear and easy to interpret. The next two or three years should see a considerable advance in our understanding of this particular problem, but physiologists' problems will even then be by no means over. When we know in the greatest detail what happens when GA_3 enters a barley aleurone cell we shall still be left to find out what its mode of action is in the many other complex responses in which it is so active and, indeed, what relationship all this work has to the functioning of gibberellins in their endogenous control of growth.

7.2 The mechanism of action of auxins

The history of studies on auxin action long antedates the sophisticated techniques now available for studies of protein and nucleic acid chemistry. Many chemical studies have been carried out on tissues treated with IAA, and it is known that a large number of chemical changes follow auxin treatment. Many of these, however, take place a relatively long time after the beginning of treatment and must certainly be regarded as secondary effects. Protein and nucleic acid synthesis occur soon after auxin treatment in many systems, but can be detected only after a lag phase of up to about an hour. In the wheat and oat coleoptile and in etiolated pea stem segments growth induced by auxin is already at a maximum by the time such synthesis can be detected. This may mean either that nucleic acid and protein synthesis are not required for the response, at least in its early stages, or that our techniques for detecting the relevant changes are not sufficiently sensitive.

Inhibitors of protein and nucleic acid synthesis do indeed affect auxin-induced growth in coleoptiles and pea stem segments. In the latter, it has been found that the bulk of the RNA produced as a result of auxin treatment is not mRNA but ribosomal RNA, though mRNA also increases. Other experiments with pea sections have shown that by the use of low concentrations of inhibitors of RNA synthesis, it is possible to inhibit such synthesis without actually inhibiting auxin-induced growth. PENNY and GALSTON (1966) suggested as a result of kinetic studies that in the early stages of auxin-induced growth some previously formed substance was used up. This explained why some auxin-induced growth took place in the absence of nucleic acid and protein synthesis. It is not surprising that the later stages of such growth should be prevented by nucleic acid synthesis inhibitors, because almost all of the many chemical changes taking place as a result of auxin treatment require new protein synthesis. Some of the enzymes involved in cell wall loosening and growth in auxin-treated tissues may increase many hours or even days after auxin treatment. Many of these facts are well discussed in a readable and informative review by GALSTON and DAVIES (1969).

There is still much discussion between those who feel that auxin action is likely to be mediated through an effect on nucleic acid and protein metabolism and those who, largely on the basis of kinetic studies, think that in some systems at least other mechanisms must be operating. Probably the strongest evidence for the second view comes from recent studies in which growth changes after auxin treatment have been followed closely or continuously. The presence of a temperature-dependent lag phase before any response is observed has always been a strong argument in favour of preliminary synthesis of some cell component before growth begins. NISSL and ZENK (1969), by means of elegant experiments with coleoptiles completely bathed in an auxin solution and observed microscopically as they responded, showed that under appropriate conditions the lag phase of growth could be reduced to zero (Fig. 7.3). If increased growth starts almost instantly on auxin treatment it is clear that the auxin effect must initially be a very direct one indeed.

Using a continuously recording photoelectric measuring device BARKLEY and EVANS (1970) carried out similar experiments with etiolated pea stem sections, and although in their case there was a minimum lag phase of about

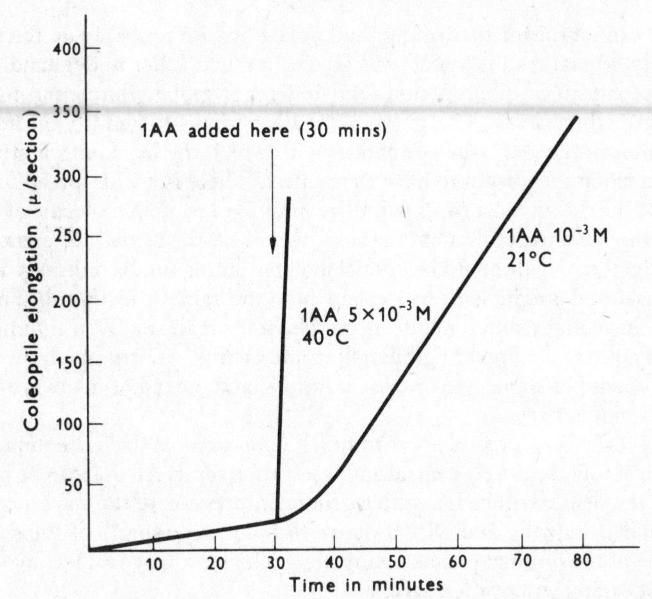

Fig. 7-3 The effect of IAA on the growth of 10 mm sections of *Avena* coleoptiles. Growth was observed under the microscope and IAA was introduced at the time marked by the arrow. Note that at higher temperatures the effect of IAA is virtually immediate. (Modified from NISSL and ZENK, 1969.)

ten minutes before any growth changes could be measured they showed that Actinomycin D did not lengthen the lag phase. This argues against RNA synthesis as a pre-requirement for the growth response. In none of these systems is there any doubt that protein synthesis is intimately involved in the response, but if it is not the very first response we are still left looking for the site and mechanism of primary auxin action. The speed of the response argues for a very accessible site of action and it may well be that attention should be turned again to the cell wall, which is known to have proteins integrally associated with it. This thought was also suggested in quite a different type of study by ANDREAE (1967). Andreae measured growth inhibition by auxin on pea root segments. He tried to establish a relationship between the growth response and a large number of cell changes likely to be associated with the response to auxin, such as the uptake of auxin from the external solution, the fate of auxin and its possible metabolites, or the formation of IAA conjugates within the cell. Of all the things measured the only one which was directly correlated with the growth response was the concentration of auxin in the external solution. This result was, to say the least, an unexpected one, and the author's conclusion was that it seemed necessary to seek a site of action actually outside the cell membrane—in fact in the cell wall.

One cannot escape the feeling that, at last, we are probably on the verge of really crucial results which will give us a much fuller understanding of the mechanism of auxin action (and indeed of gibberellin action too) in some growth systems. The greatest need is perhaps to identify the 'target' molecules in the cell—the substances with which the hormones first make contact when they begin to have their effect. There is a widespread feeling that the 'target' in most cases will turn out to be a protein molecule of some kind; the advantage of such a system would be that small differences in the 'target' molecule could be crucial in determining the second stage in the response, and might help to explain how the same growth substance is able to have many different effects on the same organism. This argument is not, of course, confined to gibberellins and auxins, as very much the same could be said of some cytokinins, inhibitors, and indeed of many synthetic growth regulators.

It has only been possible here to touch upon some of the more important aspects of this absolutely central and fascinating topic. All that can be hoped is that the reader will be left with a general impression of the ways in which the wind is blowing and with a desire to follow up aspects of the subject in some of the original papers, many of which give a very real feeling of the present excitement of the chase.

The future 8

8.1 Unresolved problems

When we look at the possible future directions of research into endogenous plant growth substances some of the gaps in our present knowledge quickly become obvious. For example, we are still a long way from having a clear idea of the exact way in which these substances do their work. It is certainly in this area that some of the most significant results can be expected in the next few years, as it is one which is engaging the attention of an increasing number of physiologists all over the world.

We also lack very detailed knowledge of the synthesis and breakdown of plant hormones. The ever-increasing sophistication of the physiological and biochemical techniques which are available for studies on complex substances will undoubtedly play a great part in enabling us to fill in, bit by bit, what amounts to a series of extremely complex jig-saw puzzles. However, the difficulties of interpretation which beset even the most careful work of this kind are very well illustrated by the studies described on the biosynthesis of IAA itself in plants. Ten years ago many plant physiologists would have felt that we were approaching the stage where we could consider IAA biosynthesis in plants to be an area of research in which really conclusive answers were only just round the corner, but these have not so far emerged. In fact set-backs like this are sometimes very good for scientists. They remind us that even the most solid-looking fact may suddenly need a complete re-appraisal in the light of new knowledge.

Another area in which we can expect some fairly substantial filling-in of gaps in our knowledge is in the study of the movement of growth substances within the plant, and the ways in which this is controlled. This involves not only the study of the long-distance transport of substances from one part of the plant to another, but also the much more difficult problem of the movement of growth substances from cell to cell over very short distances in areas of high metabolic activity, and even of the control of hormone movement within individual cells.

Some of the gaps in our present knowledge are less obvious than these, and it is certainly extremely difficult to see how some of them can be approached. A simple example here is the whole question of the ways in which the different types of plant growth substances interact with each other in the control of growth. Straightforward antagonistic relationships between two substances are not too difficult to understand or to explain, but as soon as the relationships become more complex, with synergistic interactions, or with more than two substances acting at the same time, the problems become not only practically but also conceptually much more difficult to grasp. Just because problems are difficult, however, there is no reason to suppose that they will not be tackled and eventually solved. What will undoubtedly happen during this process is that many new problems will be opened up on the way. This is one of the pleasures as well as one of the frustrations of science.

It may be worth considering at this point what new methods and research tools may become available in the foreseeable future which will enable research workers to tackle some of the unsolved problems. Some of these can be fairly easily envisaged without wandering too far into the realms of pure speculation. For example, biochemists will certainly provide us with substances which will inhibit some of the effects of plant growth substances specifically whilst allowing others to be expressed normally. This sort of approach will be invaluable in unravelling the complicated chains of events which must take place when substances with multiple properties are involved. We may also hope for substances which will inhibit the synthesis of endogenous growth substances. We have seen earlier how useful some of the inhibitors of gibberellin biosynthesis have been, and there is every reason to suppose that similarly valuable results would be obtained with inhibitors of cytokinin, ABA or ethylene biosynthesis. It is one of the paradoxes of physiological research that we can often learn more about a process by stopping it than by letting it continue normally.

It was only in 1969 that the first successful synthesis of a gibberellin was announced, but the molecules are so complex and the synthesis so difficult that it will be some time before synthetic gibberellins are at all freely available. Once this problem has been overcome a whole new field of investigation will be opened up, as we shall be in a position to do much more systematic work on the relationship between activity and chemical structure than has been possible up to now, and from this there should come a more complete understanding of how the natural gibberellins do their various jobs.

A final development to which we can confidently look forward is the increasing use of physico-chemical techniques for identifying and measuring quantities of plant growth substances extracted from plants. Great strides have already been made in this field with the developments we have already discussed, but these techniques are still only in their infancy in plant physiological studies. They must surely be regarded as pretty lusty infants, and much can be expected of them.

A question which may well be asked about the future is 'Are we going to discover new types of endogenous plant growth substance?' There is no doubt that more substances will be discovered which are similar to those we already know. This is mainly a question of technique and of time. Whether new *classes* of growth substances will emerge is altogether more problematical and we can really only speculate on it. It is certainly unwise to rule out the possibility. Many years were spent trying to explain plant growth and its control in terms of indole auxins alone before the gibberellins became widely known and it would be naïve to assert that this could not happen again. However, the groups of plant growth substances we now know offer a much more varied, flexible and understandable system of growth control than was envisaged even twenty years ago, and it is at least possible that we shall not find any more major groups of substances of such overall significance as these. This does not mean, however, that there are no more hormonal substances to be discovered. There are already reports of substances which resemble known hormones in properties though not in chemical structure. An example is phaseolic acid (Fig. 8.1), a compound isolated from bean seeds by REDEMAN *et al.* (1968) and which has gibberellin-like properties though an extremely un-gibberellin-like chemical structure. Its physiological significance is not yet known but its very existence suggests that we may well find other such chemicals in other plants, and the search for these will doubtless go on.

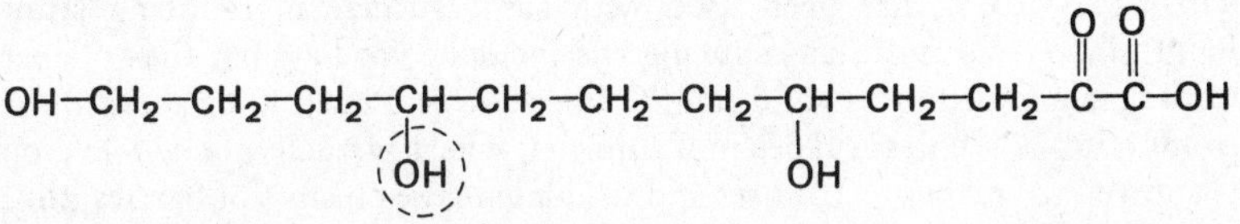

$$OH{-}CH_2{-}CH_2{-}CH_2{-}\underset{\underset{(OH)}{|}}{CH}{-}CH_2{-}CH_2{-}CH_2{-}\underset{\underset{OH}{|}}{CH}{-}CH_2{-}CH_2{-}\overset{\overset{O}{\|}}{C}{-}\overset{\overset{O}{\|}}{C}{-}OH$$

Fig. 8-1 Phaseolic acid. The position of the circled hydroxyl group is not yet absolutely certain.

It is also possible that there may be as yet undiscovered groups of hormone-like substances which have properties related to very specific physiological processes. Physiologists have long sought a hormone which would be shown to be a controlling influence in flowering, and there are other processes which may eventually be shown to have special hormonal substances as a key step in their control mechanisms. The simple fact is that we really know remarkably little about how most physiological processes are controlled in spite of all the efforts which have been made, and this remains the most crucial and certainly the most fascinating area of research in plant hormone physiology.

8.2 Applied aspects of plant growth substances

It is a curious fact that direct applications of our knowledge of endogenous plant growth substances to practical problems of controlling plant growth are relatively few. Most of the chemicals which are sold commercially and which are used for this purpose are synthetic, and have often been discovered as a result of intensive screening of many thousands of compounds in the search for useful biological activity. Compounds which have biological activity resembling indole auxins have been known for many years, and some of these such as 2,4-dichlorophenoxy acetic acid (2,4-D, Fig. 1.2, p. 3) form the basis of many weedkillers. The so-called hormone rooting compounds, used to encourage the rooting of cuttings, are also based on these chemicals, examples of which are naphthalene acetic acid (NAA) and indole

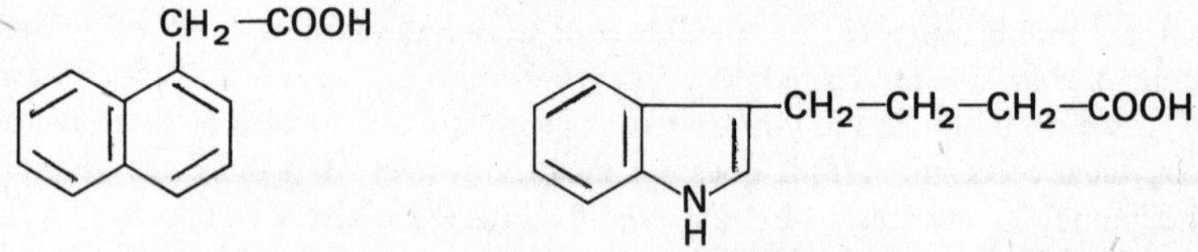

$CH_2—COOH$

$—CH_2—CH_2—CH_2—COOH$

N
H

Fig. 8-2 The synthetic growth regulators naphthalene acetic acid (left) and indole butyric acid (right).

butyric acid (IBA), (Fig. 8.2). Much more recently some synthetic growth retardants have also come to have commercial uses. Some of these are known to operate specifically by their effect of reducing gibberellin biosynthesis and thus the content of endogenous gibberellins in the plant. One of these substances, CCC, has been used with some success in reducing straw length in cereals and thus reducing the incidence of lodging, the collapse of the plants due to wind and rain. Gibberellic acid itself finds a use in the production of seedless grapes in California, where virtually the whole crop is sprayed to encourage fruit set and to elongate the clusters of berries, thus reducing the incidence of fungal rots due to the high humidity at the centre of the bunch. The same substance is used in the brewing industry to speed up the germination or malting stage of barley. Accounts of the practical uses of many of these chemicals are given by AUDUS (1959), STUART and CATHEY (1961) and MITCHELL (1966).

Table 9 Some growth and developmental processes affected by endogenous plant growth substances, or synthetic substances related to these.

Growth or developmental process	*Effect of growth substance*	*Substance used*
Stem growth	Increased	Gibberellins
	Decreased	Inhibitors
Leaf expansion	Increased	Gibberellins
Apical dominance	Enhanced	Gibberellins (often)
	Reduced	Cytokinins
		Inhibitors of apical bud growth
Senescence	Retarded	Cytokinins
	Accelerated	Ethylene
Abscission	Retarded	Some auxins
	Accelerated	Ethylene
		Abscisic acid
Dormancy	Enhanced	Some inhibitors
	Broken	Gibberellins
Flowering	Accelerated	Gibberellins (some plants)
		Ethylene (pineapples)
Rooting of cuttings	Improved	Some auxins

Compared, however, with the apparent potential of hormone-like substances in the artificial control of growth their actual use does seem a little disappointing. If we consider briefly some of the things it is possible to do with the substances we have been discussing in earlier chapters this point is perhaps made clearer. Table 9 lists some of the processes it is possible to affect either by the use of chemicals known to occur endogenously or by synthetic chemicals related to these by their properties.

Reference to the Table merely gives an indication of possibilities, some of which have already been utilized commercially while others are so far only at the experimental stage. Even apparently obvious cases of a possible practical use may sometimes turn out to be illusory. For example, gibberellic acid will cause many rosette plants to form stems and to flower in one year when their normal habit is biennial. Sugar beet is an important commercial crop which is a biennial. Much breeding work has gone into producing strains which will not flower in their first year, since such flowering causes unacceptable losses of sugar yield. As far as the breeder is concerned, however, if the plant could be made to flower in the first year this would be a great advantage in the speeding up of the breeding programme. Unfortunately the application of gibberellic acid to sugar beet does not induce flowering (for which a period of cold seems to be essential) though it causes bolting of the stem. It may be possible to find a different gibberellin, or a combination of chemical treatments which will have the desired result, but this has not so far been achieved.

As far as the use of plant growth substances in the control of crop growth is concerned the major problem is perhaps that of identifying with precision the change which we want to make. In plants grown under very intensive conditions, where normal behaviour is very well understood, it may be possible to say, for example, that stems 10% shorter, or leaf fall a week later, or flowering a week earlier would be advantageous. In many cases the desired change cannot be identified with such precision, and this presents problems to the physiologist in knowing what to try first, because it would be easy to spend half a lifetime trying the effects of all sorts of substances on the growth of a crop before discovering a really useful treatment. This problem perhaps applies especially to field crops.

It may very well be objected to the whole idea of chemical plant growth control by these methods that all the changes discussed can be equally well achieved by plant breeders, and that changes produced by them are permanent and, in the long run, cheaper than those made by chemical methods. This is quite true; no physiologist would dispute the fact that most of the long-term answers to bigger, better, more productive crop plants lie with plant breeders. Unfortunately plant breeding is a laborious and time-consuming process, and if the physiologist can manipulate the plant in desirable directions at a commercially acceptable cost until breeding programmes have had a chance to catch up he will earn the thanks not only of the richer crop producer and less hungry population, but also of the plant breeder himself, who will have been given some much needed breathing space.

Finally, in justification of all the time and effort which has been and is being put into the study of endogenous plant growth substances and their role in the control of plant growth, one may make just two points. There are few more intellectually satisfying tasks than finding out what makes a machine work; the more sophisticated the machine the more challenging the task. The growth of living cells represents the end product of the workings of a very highly sophisticated machine indeed, and the unravelling of the workings of the mechanism deserves and rewards every possible effort which can be put into it. Equally stimulating, though, is the concomitant that the more we know about the endogenous control of growth, and the greater the precision of our knowledge, the greater becomes our opportunity to manipulate growth in desirable directions. In an age of world hunger and increasing population this does not seem a bad aim for the life work of any of us.

References

ADDICOTT, F. T. and LYON, J. L. (1969). *A. Rev. Pl. Physiol.* **20**, 139–64.
ANDREAE, W. A. (1967). *Can. J. Bot.* **45**, 737–53.
AUDUS, L. J. (1959). *Plant Growth Substances*, 2nd edn., Leonard Hill, London.
BAILISS, K. W. and HILL, T. A. (1971). *Bot. Rev.* **37**, 437–79.
BARKLEY, G. M. and EVANS, M. L. (1970). *Pl. Physiol., Lancaster*, **45**, 143–7.
BEERMAN, W. and CLEVER, U. (1964). *Scient. Am.* **210**, 50–8.
BURG, S. P. and BURG, E. A. (1965). *Science, N.Y.* **148**, 1190–6.
CATALANO, M. and HILL, T. A. (1969). *Nature, Lond.* **222**, 985–6.
CHADWICK, A. V. and BURG, S. P. (1967). *Pl. Physiol., Lancaster*, **42**, 415–20.
CHADWICK, A. V. and BURG, S. P. (1970). *Pl. Physiol., Lancaster*, **45**, 192–200.
CHRISPEELS, M. J. and VARNER, J. E. (1967a). *Pl. Physiol., Lancaster*, **42**, 398–406.
CHRISPEELS, M. J. and VARNER, J. E. (1967b). *Pl. Physiol., Lancaster*, **42**, 1008–16.
CLELAND, R. E. (1969). The Gibberellins. In *The Physiology of Plant Growth and Development*. Edited by M. B. WILKINS. McGraw-Hill, New York, Toronto, Sydney, Mexico, Johannesburg and Panama.
COPPAGE, J. and HILL, T. A. (1973). *J. biol. Ed.*, **7**, No. 1, 11–18.
CORNFORTH, J. W., MILBORROW, B. V., RYBACK, G., ROTHWELL, K. and WAIN, R. L. (1966). *Nature, Lond.* **211**, 742–3.
DOMANSKI, R. and KOZLOWSKI, T. T. (1968). *Can. J. Bot.* **46**, 397–403.
FOX, J. E. (1969). The Cytokinins. In *The Physiology of Plant Growth and Development*. McGraw-Hill, New York, Toronto, Sydney, Mexico, Johannesburg and Panama.
FREELAND, P. W. (1972) *J. biol. Ed.*, **6**, No. 6, 369–375.
GALSTON, A. W. and DAVIES, P. J. (1969). *Science*, **163**, 1288–97.
GOESCHL, J. D., RAPPAPORT, L. and PRATT, H. K. (1966). *Pl. Physiol., Lancaster*, **41**, 877–85.
HELGESON, J. P. (1968). *Science, N.Y.* **161**, 974–81.
KAWASE, M. (1961). *Proc. Am. Soc. hort. Sci.* **78**, 532–44.
LANG, A. (1970). *A. Rev. Pl. Physiol.* **21**, 537–70.
LEOPOLD, A. C. (1964). *Plant Growth and Development*. McGraw-Hill, New York, San Francisco, Toronto and London.
LETHAM, D. S. (1969). *Bioscience*, **19**, 309–16.
LIBBERT, E., KAISER, W. and KUNERT, R. (1969). *Physiol. Plant.* **22**, 432–9.
MACMILLAN, J. and PRYCE, R. J. (1968). Recent studies of endogenous plant growth substances using combined gas chromatography-mass spectrometry. In *Plant Growth Regulators*. London, Society for Chemical Industry Monograph 31.
MANN, J. D. and JAWORSKI, E. G. (1970). *Planta*, **92**, 285–91.
MAPSON, L. W. (1970). *Endeavour*, **29**, 29–33.
MILBORROW, B. V. (1967). *Planta*, **76**, 93–113.
MILBORROW, B. V. (1970). *J. exp. Bot.* **21**, 17–29.
MITCHELL, J. W. (1966). *Agric. Sci. Rev.* **4**, 27–36.
MITCHELL, J. W. and LIVINGSTONE, G. A. (Editors) (1968). *Methods of Studying Plant Hormones and Growth-Regulating Substances*. Agriculture Handbook 336, Agricultural Research Service U.S. Department of Agriculture, Washington, D.C.

NISSL, D. and ZENK, M. H. (1969). *Planta*, **89**, 323–41.
NITSCH, J. P. (1950). *Amer. J. Bot.* **37**, 211–15.
OSBORNE, D. J. (1965). *J. Sci. Fd. Agric.* **16**, 1–13.
OSBORNE, D. J. (1968). Ethylene as a Plant Hormone. In *Plant Growth Regulators*. London, Society for Chemical Industry Monograph 31.
OVERBEEK, J. VAN. (1966). *Science, N.Y.* **152**, 721–31.
PALEG, L. G. (1965). *A. Rev. Pl. Physiol.* **16**, 291–322.
PENNY, P. and GALSTON, A. W. (1966). *Amer. J. Bot.* **53**, 1–7.
PHILLIPS, I. D. J. (1971). *Introduction to the Physiology and Biochemistry of Plant Growth Hormones*. McGraw-Hill, New York, London, Toronto, Sydney.
PRATT, H. D. and GOESCHL, J. D. (1969). *A. Rev. Pl. Physiol.* **20**, 541–84.
PRYCE, R. J. (1971). *Planta*, **97**, 354–7.
RADLEY, M. (1956). *Nature, Lond.* **178**, 1070–1.
REDEMAN, C. T., RAPPAPORT, L. and THOMPSON, R. H. (1968). Phaseolic acid: a new plant growth regulator from bean seeds. In *The Physiology and Biochemistry of Plant Growth Substances*. Edited by F. WIGHTMAN and G. SETTERFIELD. Runge Press, Ottawa.
SETH, A. and WAREING, P. F. (1967). *J. exp. Bot.* **18**, 65–77.
SITTON, D., ITAI, C. and KENDE, H. (1967). *Planta*, **73**, 296–300.
SKOOG, F. and ARMSTRONG, D. J. (1970). *A. Rev. Pl. Physiol.* **21**, 359–84.
STEEVES, T. A. and BRIGGS, W. R. (1960). *J. exp. Bot.* **11**, 45–67.
STOWE, B. B., STODOLA, F. H., HAYASHI, T. and BRIAN, P. W. (1961). The Early History of Gibberellin Research. In *Plant Growth Regulation*. Edited by R. M. KLEIN. Iowa State Press, Ames, Iowa.
STUART, N. W. and CATHEY, H. M. (1961). *A. Rev. Pl. Physiol.* **12**, 369–94.
THIMANN, K. V. (1969). The Auxins. In *The Physiology of Plant Growth and Development*. Edited by M. B. WILKINS. McGraw-Hill, New York, Toronto, Sydney, Mexico, Johannesburg and Panama.
VALIO, I. F. M., BURDEN, R. S. and SCHWABE, W. W. (1969). *Nature, Lond.* **223**, 1176–8.
WAREING, P. F. and PHILLIPS, I. D. J. (1970). *The Control of Growth and Differentiation in Plants*. Pergamon Press, Oxford, New York, Toronto, Sydney, Braunschweig.
WAREING, P. F. and RYBACK, G. (1970). *Endeavour*, **107**, 84–8.
WENT, F. W. and THIMANN, K. V. (1937). *Phytohormones*. Macmillan, New York.
WEST, C. A. and PHINNEY, B. O. (1956). *Pl. Physiol. Lancaster*, **31**, Suppl. xx .
WILKINS, M. B. (Editor), (1969). *The Physiology of Plant Growth and Development*. McGraw-Hill, New York, Toronto, Sydney, Mexico, Johannesburg and Panama.
WRIGHT, S. T. C. (1961). *Nature, Lond.*, **190**, 699–700.
WRIGHT, S. T. C. (1968). Multiple and Sequential Roles of Plant Growth Regulators. In *The Physiology and Biochemistry of Plant Growth Substances*. Edited by F. WIGHTMAN and G. SETTERFIELD. Runge Press, Ottawa.
WRIGHT, S. T. C. and HIRON, R. W. P. (1969). *Nature, Lond.* **224**, 719–20.